FUNGI
DECODED

CLARKSON POTTER/PUBLISHERS
An imprint of the Crown Publishing Group
A division of Penguin Random House LLC
1745 Broadway
New York, NY 10019
clarksonpotter.com
penguinrandomhouse.com

2025 Clarkson Potter/Publishers Trade Paperback Edition

ISBN 979-8-217-03396-6
Ebook ISBN 979-8-217-03399-7

Published in the United Kingdom as FUNGI DECODED: ENCOUNTERS WITH THIRTY FUNGI by UniPress Books Ltd., London.

Manufactured in China

1 2 3 4 5 6 7 8 9 10

Front cover: Shutterstock/Phusita Walklate.
Back cover (right), back flap: Shutterstock/Anitapol.

The authorized representative in the EU for product safety and compliance is Penguin Random House Ireland, Morrison Chambers, 32 Nassau Street, Dublin D02 YH68, Ireland, https://eu-contact.penguin.ie.

MIX
Paper | Supporting
responsible forestry
FSC
www.fsc.org
FSC™ C005748

FUNGI
DECODED

ENCOUNTERS WITH THIRTY FUNGI

BRITT A.
BUNYARD

POTTER

CONTENTS

2

The Secret World of Fungi

3

A Ring of Thirty Fungi

1

Introducing
Fungi

INTRODUCTION

*Fungi are without doubt
the most important link
among all life on the planet.*

Fungi have evolved the ability to utilize almost anything as a food source; they trick plants—and animals—into doing their bidding. They effectively poison, trap, kill, enslave, and zombify other living organisms. And they are without doubt the most important link among all life on the planet.

Neither plants nor animals, fungi form their own kingdom of living things. Although quite different from us, they are more like animals than any other life-form. This book aims to introduce some of the more charismatic species from the mysterious kingdom of fungi.

Most species of fungi we will probably never see nor get to know; most are microbes and too small to see with the unaided eye. However, some fungi produce reproductive structures (mushrooms) that are large enough to be seen, and these "macrofungi" will be the main focus of this book.

Unlike plants, which can create their own food by harnessing the energy of sunlight, fungi have to take energy and nutrients from other organisms (what's called a heterotrophic lifestyle).

Most fungi survive by doing one of three things in the environment. Many fungi are saprotrophic, meaning they break down dead matter from other organisms in order to live. Others get what they need by parasitizing a living host: many animal and plant diseases are caused by molds and other fungi; and others live as partners with other organisms, especially plants. Some fungi are harmful to their hosts; others may be toxic if eaten. But the more we learn about them, the more we see that the healthy function of the natural world around us is closely tied to them.

PHEASANT'S BACK: This large mushroom is one of a group called saprobes, usually found on rotting logs and tree stumps.

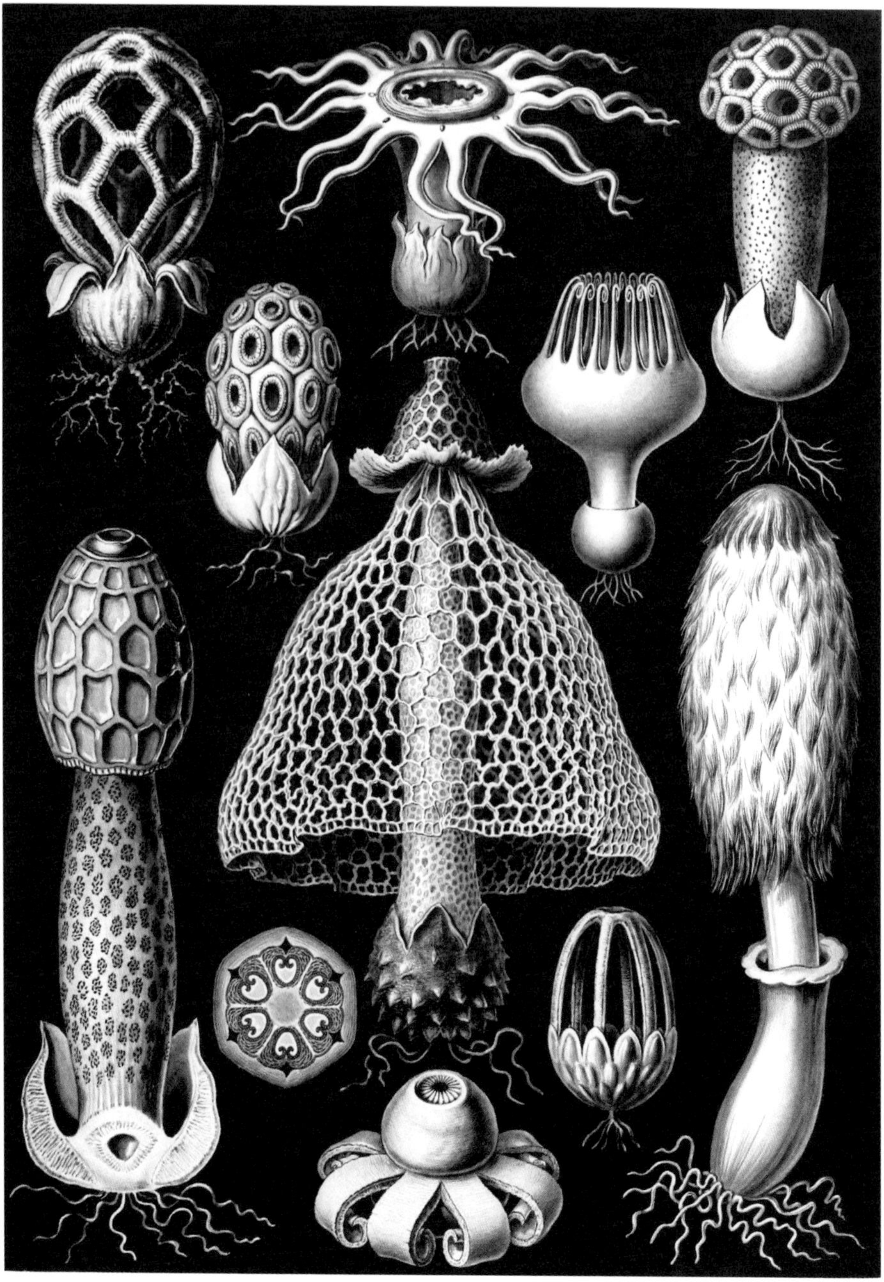

Fungi grow in a wide variety of habitats, and most,
if not all, have an important role in the environment.
Many species are specific about their food source and
will be found near certain kinds of trees. Some prefer dry
areas such as deserts or beaches. Many fungi are important
as decomposing organisms, aiding in the breakdown
of wood, leaves, and other organic debris in woodlands,
grasslands, and even urban settings where woody mulch
is used. This results in the recycling of essential nutrients.

If you are fascinated by fungi but perplexed by their
myriad forms and colors, this book is for you. With careful
observations of shapes, colors, odors, and so on, along with
noting the habitat in which they are found, it is possible
to get to know and better understand them.

THE LANGUAGE OF MUSHROOMS

Fungi are consummate communicators of the natural world. We now know that underground networks of fungal threads (called mycelia) form connections between trees and other plants, stretching across huge distances. It seems likely that these networks help ecosystems to stay healthy by communicating information about threats, where to find food, and when it's time to reproduce. By learning how to decode what fungi are "saying," we can better understand and appreciate them.

Learning the "language" of organisms so very different from us may seem like a daunting task, but it's not impossible. If you have ever been in a situation where two people were communicating in a language foreign to you, it may have seemed as if they were using a strange code. And yet, you might have been able to figure out what they were saying to each other by observing how they were saying it, the situation, or the time and place. We humans can apply the same process to understanding other living things.

The study of living organisms is called biology; the study, specifically, of fungi is called mycology. Getting to know any group of living organisms mostly involves observation,

LEFT
FUNGAL CONNECTIONS:
Hypholoma lateritium fruit in reddish clusters from wood.

MIDDLE LEFT
FRUITING BODIES:
Hypholoma fasciculare have a yellowish color, produce black spores, and fruit in clusters from woody debris on soil. The solitary banded cort (*Cortinarius armillatus*) features a bright orange band on the stalk and produces rust-brown spores.

BELOW
FAR-REACHING FUNGI:
The turkey tail grows in clusters of thin brackets on wood, all over the world. It is widely revered for its medicinal properties.

There's much you can learn about fungi in your own backyard.

with a bit of investigation. Observation may include noting the organism's appearance (size, shape, color), odor, taste, habitat, behaviors, and interactions with its surroundings. Decoding the life cycle of an organism also involves understanding its reproduction, growth, and development.

Carrying out these kinds of close observations can be fascinating, but the good news is, you don't have to start from scratch. Books and published information can help to decode the evolution and taxonomy of living things. Taxonomy, simply, is the classification of all organisms into groups. It, too, comes with its own language, and in time, this will become part of your lexicon. Taxonomy enables us to better understand the visible characteristics of an organism, as well as to predict things about that organism that we haven't yet seen first-hand.

The language of mycology

One of the features of taxonomy is that every species has its own scientific (Latin) name. If you find yourself hanging out with mushroom enthusiasts (mycophiles), you will likely hear multiple ways to pronounce the same scientific name. Inevitably, you will witness self-proclaimed "experts" going so far as to argue over which way is correct. Do not be intimidated: they are undoubtedly both wrong.

For most organisms, Latin is the basis for scientific names; mycologists are extra fiendish and also use Greek. Mycological names are often a combination of Latin *and* Greek. I love to hear my Italian colleagues rattle off the scientific names of fungi—they are likely closest to the correct pronunciation.

Most of us, however, will mispronounce names. And that's okay. A useful tip: with scientific names of organisms and structures, you should pronounce each vowel individually, and there may be a person's name mixed in to add to the confusion. For example, *Boletus smithii*, a species of *Boletus* mushroom named for the mycologist Alexander H. Smith, is pronounced *Bo-lee-tus smith-e-aye*. The two that really give me pause, and that no one should mispronounce, are "oospore" and "zoospore." "Oospore" is pronounced *oh-oh-spore*, not *oo-spore*; likewise, "zoospore" is pronounced *zo-o-spore*, not *zoo-spore*. Similarly, the study of animals is zoology, pronounced *zo-ol-o-gy*, not *zoo-ol-o-gy*. (The latter, presumably, is the study of zoos and would actually be spelled "zooology." Which looks ridiculous.)

Observational anatomy

Recently, scientists have developed tools that allow us to decode the genomes of organisms (and therefore, all the DNA that codes for everything about that organism), which furthers our understanding of taxonomy. But there is so much about fungi that you can learn from simple observations without the use of a microscope or test tubes, and often right in your own backyard or nearby woods.

OVERLEAF
A COLORFUL PALETTE:
In addition to many forms, mushrooms
come in just about every color.

What makes a fungus?

Mycologists believe that there may be around a million species of fungi on the planet, of which only 100,000 have been studied. So 90 percent of species remain unknown.

One of the unique features of fungi is that their cells have a different construction from those of plants and animals. Fungus cells have walls constructed of substances called chitinous carbohydrates, long molecular chains somewhat similar to the material that makes up an insect's exoskeleton.

Plant cell walls are constructed of stuff called cellulosic carbohydrates, cellulose being the main component of plant fiber. Animal cells have a cell membrane only, no cell walls.

The body of the fungus is called a mycelium, and it is made up of a network of thread-like cells called hyphae (singular: hypha). Mycelia largely grow underground, and hyphae are frequently too small to be seen with the naked eye.

The more familiar part of the fungus is the mushroom. Mushrooms are the above-ground extensions of fungi, which grow to spread spores (the method by which fungi reproduce).

It is likely that 90 percent of fungi species have yet to be discovered.

MUSHROOM
MORPHOLOGY:
Turn over a mushroom and
you may see gills. Besides
their elegance, gills—their
attachment and appearance—
are useful in identification.

The parts of a mushroom often include a cap, stem (stalk),
and the hymenium, where the spores are produced, which
comes in many forms, such as gills, pores, and spines.

The cap (or pileus), at the top of the mushroom, is usually
the first feature to be studied. When immature, the cap can
be round, conical, bell-shaped, or convex. As the mushroom
matures, the cap often flattens out. Other caps can become
vase-shaped or even have a knob on the top. Look closely
at the surface: caps can be dry, wet, or sticky. There are also
many textures that help differentiate one from another.

When a mushroom is turned over, gills, pores, or spines
will be seen. Gills appear as long, thin blades running
from the edge of the cap to the stalk. Spores are launched
from the surface of the gills. Pores, on the other hand,
are like long tubes running through the underside of
the fruiting body. They appear as small holes, from

INSIDE AND OUT:
The exterior of a mushroom is key to identification. But the interior can be useful too; when sectioned, it reveals interesting details.

which the spores are released, and hang from the underside of the cap. Mushrooms with spines are a feature of the tooth fungi; those with a stalk, cap, and gills are called agarics; and those with a stalk, cap, and pores are called boletes.

The stalk is the structure that supports the cap. Stalks don't always have to be in the center of the cap; they can be off-center, or eccentric. Stalks can be long or very short, or there may be no stalk at all in the case of sessile forms. The bottom of the stalk, at the ground level or below, can be round like a bulb or not. When cut open, the inside can be firm, hollow, or spongy—texture is important in identification. Remains of other parts of the mushroom (such as the volva or partial veil) can be seen on the stalk.

The veil is a thin tissue that covers some part of the immature mushroom. As the mushroom begins to mature, the veil breaks apart and pieces of it can be seen on the cap and stalk. A universal veil covers the entire mushroom, whereas a partial veil covers only the gills. On a mature mushroom, remnants of the universal veil may remain on the top of the cap as warts or a single patch.

At the base of the stalk there may be scales or rings of tissue, or a large cup or boot of tissue called a volva. On a mature mushroom, remnants of the partial veil may remain as an annulus or ring on the stalk.

Spores and spore prints

While often pretty to look at, fungi produce mushrooms entirely for the purpose of reproduction. The end result— the reproductive propagules—are spores, which many people assume are analogous to seeds produced by plants. (They are actually more like plant pollen, but more on that later.)

We cannot see individual spores with the naked eye—they are far too small. However, if enough of the spores drop from a mushroom and pile up (termed a spore print), you will observe that the spore color may vary from one group of mushrooms to another and be helpful in identification.

Some colors are common, such as white—many groups of agarics have white spores, and many boletes have brown spores. In contrast, only a few groups of mushrooms have pink, and only one species has green. To check the color of spores, it is easy to take a spore print. Simply place a mushroom cap on paper (or suspend it just above), cover the mushroom with a bowl (creating a chamber that will keep it humid), and leave for a few hours or overnight.

MORPHOLOGY MATTERS:
Many stalked mushrooms
growing on wood are
polypores, but look closely—
sometimes they have gills,
like these *Galerina marginalis*;
while the shaggy *Polyporus
squamosus* has pores, enabling
easy identification.

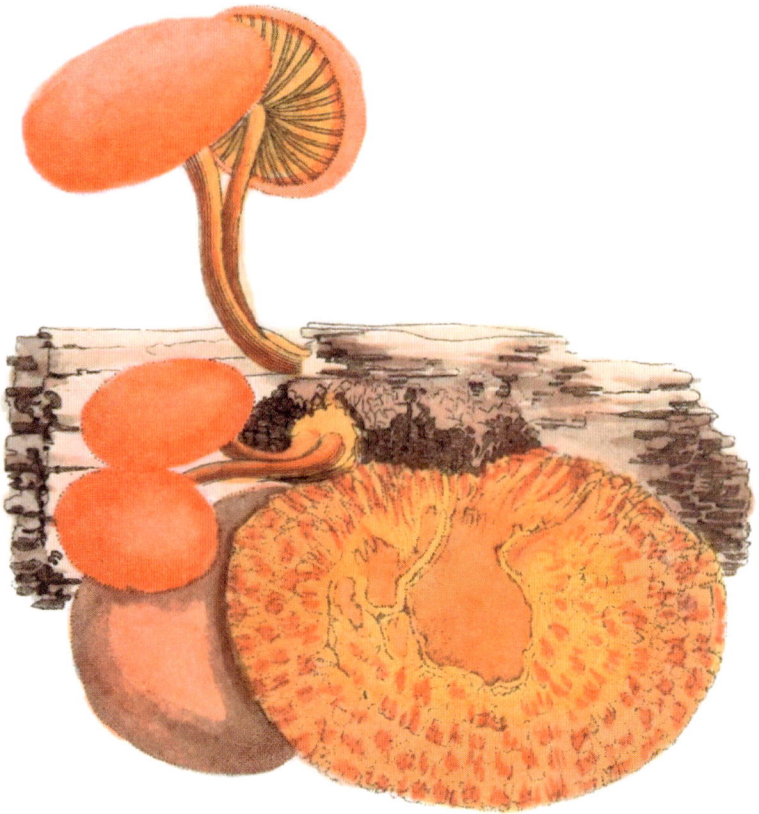

SIMILAR BUT DIFFERENT:
Toxic *Amanita pantherina*,
right, and edible *Amanita*
rubescens ("the blusher"), left,
are both brown mushrooms
but the blusher takes its name
from the reddening color that
develops on the stem.

2

The
Secret World
of Fungi

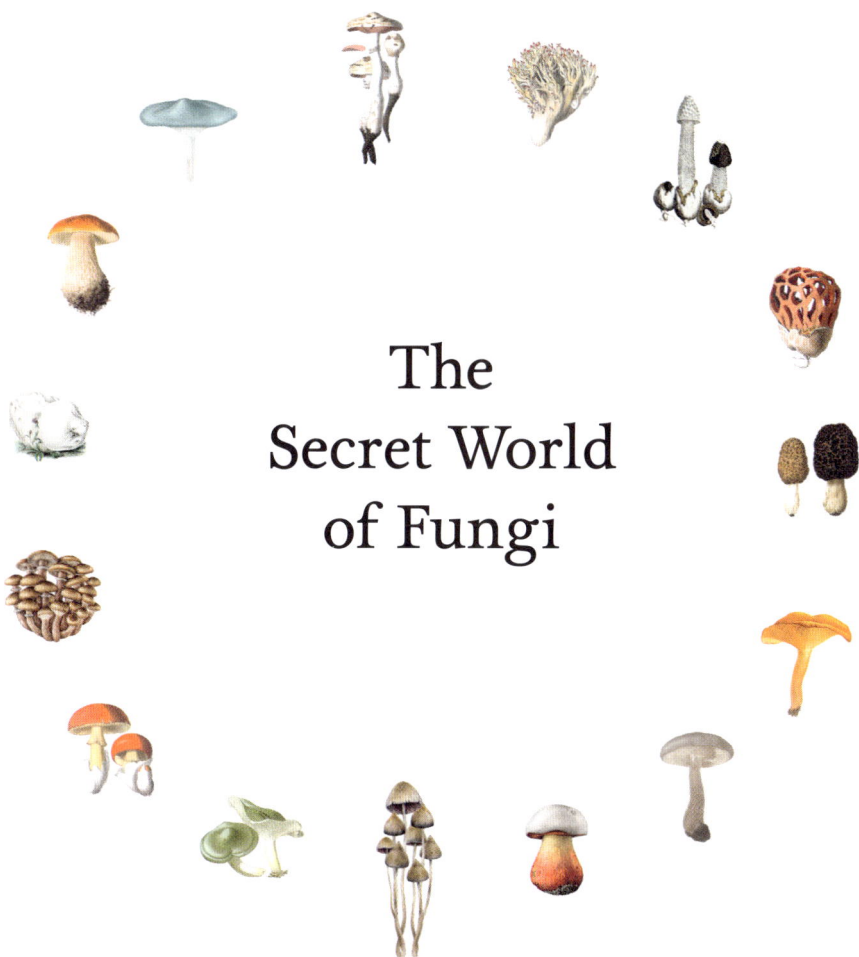

FOLLOWING THE HIDDEN THREADS

Fungi do things and live in ways that seem very foreign to us, even otherworldly. Even so, they are crucially important to all life on the planet, as we are just beginning to realize. While most are tiny and go unnoticed, some fungi produce mushrooms large enough to spot while walking through the woods. But the mushroom is just the mycological tip of the iceberg. The main bodies of the fungi remain hidden. What are they doing, and how can we come to understand them better?

Fungi comprise an entire kingdom of life. Just as members of the animal or plant kingdoms are very different from one another, so too are members of the fungal kingdom. Their ways of obtaining nutrition and their defense mechanisms, genetics, reproduction, and communication are all very different from the familiar behaviors of animals and plants.

There are three main styles of fungi growth: most fungi grow as tiny threads called multicellular filamentous hyphae; some of these fungi can also grow as a single-celled yeast form; and a few fungi favor growing (or only grow) as a single-celled yeast form.

The mushroom is just the mycological tip of the iceberg. The main bodies of fungi remain hidden.

DIVERSE HABITATS:
A cluster of fragile mushrooms signals when a fungus is rotting a log, but main bodies of the fungi remain hidden inside, growing year round.

PREVIOUS
GARISH COLORS:
Many mushrooms are found
in drab shades of brown or
white, but not all.

BELOW
NATURAL PARTNERS:
Hygrophorus russula is native to
North America and Europe,
and is generally found under
oak or pine trees.

Hyphae are long, cylindrical, branched cells. An individual hypha is approximately one twenty-fifth the diameter of a human hair. In the vast majority of fungi, hyphae are divided by septa, a sort of wall. As fungi grow, hyphae mass together to form tissue known as a mycelium; mushrooms such as boletes and chanterelles, as well as tough woody polypores, are made of hyphae.

An individual hypha (thread of fungus) is approximately one twenty-fifth the diameter of a human hair.

RIGHT
IMMATURE FORM:
When immature, *Chlorophyllum rachodes* looks like a drumstick with a long stalk and closed cap.

CENTER
MUSHROOM DIVERSITY:
Cortinarius violaceus is a striking purple color.

FAR RIGHT
MATURE FORM:
When mature, the cap of *Chlorophyllum rachodes* opens and becomes shaggy.

FALSE FUNGI

What is the key that makes a fungus a fungus? How can you tell the pretenders from the real thing?

Several groups of organisms—slime molds, "water molds," some bacteria, some plants, and even some invertebrate animals—resemble fungi. So how can you tell the pretenders from the real thing?

Myxomycetes, or slime molds, ooze along in thin streams, and sometimes even put up tiny fruiting bodies, like microscopic mushrooms. They are, however, protozoans—specifically, amoebae—so not fungi at all.

Oomycetes, or water molds, were long thought to be fungi; their hyphae are all but indistinguishable. Oomycetes are actually a kind of micro-organism called protists, with cell walls based on cellulose (not unlike plants).

Some parasitic plants send threads into their host's tissues, making them look very much like parasitic fungi. Some algae form in chains, but grow in a different way from the hyphae of fungi. Plants and algae both photosynthesize, absorbing sunlight to provide energy.

Some fungi even resemble the motile (moving) cells of creatures from other kingdoms, with a "zoosporic" stage that is capable of creeping around! So what is the key that makes a fungus a fungus?

The secret is in the cells

A key chemical component to the body of all fungi is a substance called chitin, which is made of polymers (chains) of the carbohydrate N-acetylglucosamine, a derivative of glucose. Chitin affords strength and flexibility to fungal hyphae. As we do not produce the enzymes to degrade chitin, chitinases, the chitin we ingest when we consume mushrooms and other fungi passes through us as fiber, much in the same way plant cellulose does.

One other note while on the topic of food and digestion: fungi also produce glycogen as an energy storage molecule. Glycogen is found only in fungi and animals.

THE FIRST MYCOLOGISTS

Fungi have always puzzled and fascinated people. Their strange shapes and colors, their potential as food, their psychedelic effects, and of course their dangerous toxins have made them objects of mystery and magic in folklore across the world.

RECOGNIZING MUSHROOMS:
Pliny the Elder was one of the first to decode mushrooms, identifying *Amanita muscaria*, fly agaric.

The scientific study of fungi also has a long history, dating back at least to ancient Rome. Pliny the Elder (23–79 CE), a native of Como, Italy, was a scientist, a sailor, and one of the greatest botanists of all time. He died in the eruption of Mount Vesuvius that buried Pompeii and Herculaneum. In his thirty-seven-volume *Historia Naturalis*, he described how *Amanita muscaria* grew from "eggs" or ovules, where his predecessors thought fungi arose from all manner of mystical sources, including vapors, spontaneous generation, fermentation of the soil, and woodland spirits.

In later centuries, scientific ideas around the categorization of living things, including fungi, spread more widely. Pier Antonio Micheli (1679–1737) is considered the founder of modern mycology and was the author of several key mycological discoveries. His passion for botany began while he worked as an apprentice bookbinder, where he was exposed to botanical works by Mattioli, Boccone, and other Italian greats. This passion was enhanced by the teachings of the monks of Vallombrosa and by his exploration of the forests scattered throughout Tuscany.

At age twenty-seven, Micheli became court botanist to the Grand Duke Cosimo III of Tuscany, with an annual pension and tasked with providing plants for the botanical gardens of Tuscany. Micheli's statue stands in the world-famous Uffizi Gallery in Florence, sharing the limelight with the likes of Dante, Galileo Galilei, Leonardo da Vinci, and Niccolò Machiavelli. (It seems the Italians really love their mushrooms and their mycologists!)

MASTERS OF MODERN MYCOLOGY

DEATH AND DECAY:
Polypore fungi are efficient
at breaking down and
degrading dead wood, but
some species begin their
attack on living trees.

By the early nineteenth century, mycology was a fully recognized science, completely independent of botany. Fungi were firmly established as the fifth kingdom of living beings.

It was the great South African mycologist Christiaan Hendrik Persoon (1761–1836) who brought Micheli's work to the fore. Persoon's principal work was *Synopsis Methodica Fungorum* (1801), which explored for the first time the concept that the part referred to as "fungus" was really just one part of a much more complex organism.

By the mid-nineteenth century, it had become essential to find a new classification system for fungi based on unified characteristics. Elias Magnus Fries (1794–1878) is credited with developing such a system. His taxonomic system proved so useful and was so widely accepted that, even today, it continues to be an indispensable aid for anyone preparing to study fungi.

Whether a taxonomist or merely hobbyist, you may be familiar with the Friesian system for grouping fungi. It's how nearly every mushroom field guide is arranged— white-spored mushrooms grouped together, dark-spored mushrooms, boletes, polypores, etc. It just feels natural to search for the mushroom you are trying to identify using this system. While it may not be entirely correct, evolutionarily speaking, it remains an efficient way to group mushrooms for the purposes of identification.

ABOVE LEFT
SCALY SAWGILL:
Neolentinus lepideus is recognizable by its sawtooth gills. Also known as the "train wrecker" for its ability to degrade creosote-preserved wooden railroad ties.

ABOVE
EARTHSTAR:
Some puffballs have a thick rind or perideum. When mature, this covering opens and becomes leg-like rays that hoist the mushroom above its substrate.

LEFT
WEBCAP:
Species of *Cortinarius* feature a cobwebby partial veil that covers the gills at immaturity; as the mushroom matures and the cap opens, the veil falls away to allow the rust-brown spores to be released.

RIGHT
AMETHYST DECEIVER:
Laccaria amethystina is
a beautiful purple when
young, but the color fades as
it matures, leaving it looking
very similar to other common
species of the genus.

BELOW
AGARIKON:
Larcifomes (*Fomitopsis*)
officinalis is a perennial
woodrot polypore. It is also
called the "quinine conk"
for its very bitter taste.

BELOW RIGHT
BLEEDING TOOTH:
Hydnellum peckii is a stunning
tooth mushroom with a grisly
name, but is also known as
"strawberries and cream"
for its beautiful red and
white coloration.

FUNGUS FAMILIES

The defining feature of fungus classes? How they reproduce.

Thanks to the work of mycologists over the centuries, fungi are classified into four major groups. The defining feature? How they reproduce. This system gives us four classes of fungi: chytridiomycetes, zygomycetes, basidiomycetes, and ascomycetes.

The formal class (phylum) names are capitalized: Chytridiomycota, Glomeromycota, Basidiomycota, and Ascomycota. "Zygomycota" is now seen as a sort of artificial group of fungi that share some similarities but don't have a common evolutionary ancestor. (That's why you'll often see it written in quotes.)

Basidiomycete and ascomycete fungi are often referred to as the "higher" fungi, meaning they evolved more recently than other groups. These are the most commonly known groups, as the larger, showier mushrooms are mostly basidiomycetes and a few ascomycetes.

This scheme comes with a warning: modern genetic analysis has shown us that the evolution of fungi is rather more complicated than the historical taxonomists thought. In fact, not all scientists agree on the taxonomy for some of the oddball fungus groups!

Another complication is that the four main classes are based on different forms of sexual reproduction—that is, producing offspring from the joining of male and female cells. However, sometimes fungi reproduce asexually; for many fungi, asexual reproduction is their only means of reproduction. Asexual fungi may make copies of themselves, sometimes as asexual spores, or conidia; for others it may be less sophisticated—the individual hypha may simply fragment, with the broken individuals beginning new colonies. But asexual fungi present a problem: if fungi are classified on the basis of how they reproduce sexually, then how do we classify asexual forms? In fact, a great many important fungi fall into this category; many of these cause damage to our crops, rot our stored foods, or cause diseases.

Nevertheless, even though it's somewhat simplified, the division into four classes is still a useful system for understanding what these are and how they reproduce.

BRITTLE GILLS:
Although they are known for their
bright colors of red, yellow, and
green, *Russula* species are known
as "brittle gills" for their crisp,
crumbly texture.

BIG LAUGHING GYM:
Gymnopilus spectabilis and
close relatives are indeed
spectacular to see but are also
known for their psychotropic
properties.

MOVERS AND PUPPET MASTERS

Glomeromycota are likely the puppet masters of all life on the planet.

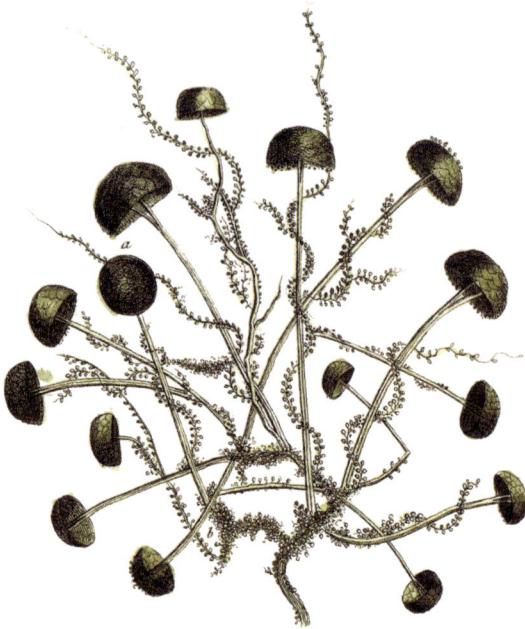

COENOCYTIC FUNGI: The earliest fungi lack many features of macrofungi more familiar to us. These fungi have no septa to divide hyphae into cells and no elaborate fruit bodies.

Two classes of fungi—Chytridiomycetes and Zygomycetes—evolved earlier than the others, and are known as "lower" fungi. They tend not to have large, colorful mushrooms, but they do have their own surprising tricks.

Chytrids are the only fungi that can actually move around. They produce asexual spores that are propelled by tiny, tail-like organs called flagella. Found worldwide, chytrids are mostly saprotrophs, which feed on rotting matter, although some species are parasites. Chytrids are linked to the worldwide die-off of amphibians.

Our next group, the zygomycetes, is a mixture of fungi grouped together by virtue of having aseptate hyphae—meaning cells in their threads are not separated. Many produce a stage in reproduction called a zygospore. Some well-known examples include black bread mold (*Rhizopus stolonifer*) and the *Pilobolus* species—the hat thrower—capable of ejecting spores great distances (see page 59).

Glomeralean (also spelled "glomalean") fungi were once thought to be part of the zygomycetes but have now been elevated to their own phylum, the Glomeromycota. These fungi are poorly known, as few have been seen or cultured. Few, if any, have sexual reproduction; they form no obvious fruit bodies; some form clusters of asexual spores . . . and that's about all we know. Except that that they grow in partnership with most plants and are likely the puppet masters of all life on the planet! (Research is ongoing, so stay tuned as this story unfolds.)

THE MOST FAMILIAR FUNGI

Most of the fungi you are likely to find come from the basidiomycete and ascomycete groups, or "higher" fungi. These are the most recently evolved of all fungi and share a common ancestor.

The ascomycetes are the largest group of fungi. They are known as sac fungi, because they produce spores in a special sac-like structure called an ascus. In addition to being the largest group, the ascomycetes are the most diverse, living as saprotrophs (feeding on decaying matter), parasites, and mutualistic symbionts (meaning partnered with other organisms). This group includes morels, truffles, and yeasts.

These are the fungi that are arguably the most fascinating and well-known.

The basidiomycetes include most of the mushrooms familiar to everyone. They are also called club fungi, because they produce spores on club-like stalks called basidia. Mushrooms are the reproductive structures, and the basidia are found on the fertile surface (hymenium): gills and pores, as well as other structures. This is a diverse group living as saprotrophs, parasites, and mutualistic symbionts. These are the fungi that are arguably the most fascinating and well-known, and therefore of greatest interest to this book.

MUSHROOMS:
Fungi are classified by how they reproduce. Most familiar to us are the mushrooms, the reproductive structures of "higher" fungi.

FUNGI REPRODUCTION: IT'S COMPLICATED

If you could sum up fungal reproduction in one word, it would have to be "complicated." Animal reproduction is straightforward by comparison. Firstly, there are only two genders involved for animals; and secondly, almost all animals reproduce exclusively by sexual means. Fungi, on the other hand, reproduce by just about any way imaginable. Most, if not all, fungi can reproduce asexually, but sexual reproduction is also common, especially in the species familiar as mushrooms.

MYRIAD FORMS:
Mushrooms come in many forms; agarics feature a stalk and cap with gills underneath. Coral mushrooms are club-like and may be branched—some highly branched and stalked.

ALL SHAPES AND SIZES: Hydnoids feature a stalk and cap with teeth underneath (left). Polypores feature pores underneath their caps and some may be stalked (right).

Sexy fungi

Whether fungi or not, sexual reproduction in all organisms involves three events. The first is where two haploid cells (cells with a single set of chromosomes, or sex cells) fuse together. This means the cell nuclei (the part containing the genetic information) share the same cytoplasm (the fluid that fills the cell). This stage is known as "plasmogamy" in fungi.

The second event is nuclear fusion (no, not the explosive kind), where the two nuclei join to create a diploid cell (meaning a cell with two sets of chromosomes). This is known as "karyogamy" in fungi.

The third event is a type of cell division called meiosis, which results in nuclei that will eventually become the offspring of the parent organisms.

Sexually reproducing fungi feature these three key events, but the stages may be separated in time or happen in close succession. Some fungi are self-fertile (homothallic), while many are heterothallic and must reproduce with another

MUSHROOM MORPHOLOGY:
One of these mushrooms
produces spores from icicle-like
teeth, the other from pores
underneath the cap.

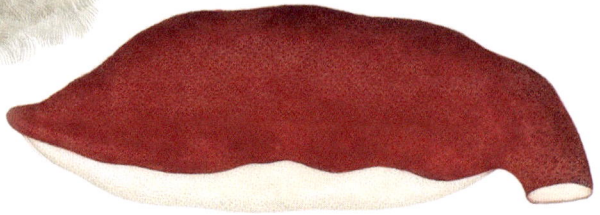

individual. For these fungi, sexual reproduction is
governed by compatibility or mating type genes. Mating
type genes allow a fungus to mate with all members of
the same species except those of the same mating type—
essentially a mechanism to avoid inbreeding. Many fungi
have just two or four mating types; on the extreme end,
Schizophyllum has 28,000 mating types.

Most fungi exhibit a life cycle with both a haploid and
diploid phase; for many, each cell contains two haploid
nuclei (one from each parent), which is known as a dikaryon.
Basidiomycetes produce haploid spores (basidiospores) that
germinate and must fuse (plasmogamy) with an opposite
mating type hypha quickly in order to restore the dikaryon
state and a full complement of necessary genes. The fungus
can live as a dikaryon for a long time; only at the time
of sexual reproduction does karyogamy take place (in the
basidium) of the fruit body. Karyogamy and the diploid state
is quickly followed by meiosis and haploid spore production.

Un-sexy fungi

The other form of reproduction, asexual reproduction, involves creating clones of the parent, and fungi can do this through a number of different mechanisms. The most common is simple mycelial fragmentation: chains of small cells from the hyphae or asexual propagules called conidia (sometimes referred to as "spores") split off and grow into a new organism.

Because asexual propagules are clones of the parent, genetic variation is reduced, but that is not to say that anamorphic fungi are not evolutionarily successful. Quite the contrary. Many of our most common and widespread fungi reproduce asexually; many of our most economically important fungal pathogens and most of the common molds spoiling our foods at home are anamorphs.

A clear advantage to asexual reproduction is that it is much more energy efficient to reproduce without having to produce large masses of tissue for fruit bodies. Most, if not all, fungi can reproduce asexually. Maybe this is the primary way fungi spread—we just don't know. Asexual reproductive structures are minute and difficult to spot in nature, so they may be more common than we realize.

Asexual reproduction in fungi is hard to spot, so it may be more common than we realize.

SPREADING THE SPORES

Where plants spread seeds, fungi spread spores, or reproductive propagules if you want to sound more technical. (Even better, in the case of spores produced asexually: conidia). We know that spore spreading works because fungi are everywhere!

You might think of spores dispersing in a passive fashion, wafting away from the fruit body on gentle air currents. It is true that spores drift on the air once they separate from the parent, but the initial release is often quite active—even explosive!

The spore-producing surface (hymenium) is often shaped to dramatically increase its surface area, giving more space for spore production. Fruit bodies may be convoluted, ribbed, gilled, branched, or covered with tubes, though some are simply a single smooth club.

SPORE SUPPORT:
Ascomycetes like cordyceps species produce spores within perithecia chambers supported on sterile stalks called stroma.

Spore squirters

Ascomycete fungi have a style of spore release that is often likened to a squirt gun. The spores are formed within a long, sac-like pouch called an ascus. Asci line the spore-producing surfaces of cup mushrooms such as *Morchella*, *Helvella*, or *Chlorociboria*. In other varieties, they may be within chambers such as perithecia, themselves hidden within stromal (or supportive) tissue as with *Cordyceps*, *Claviceps*, and *Xylaria*, to name a few examples.

As the fruit body matures, liquid flows into the ascus, causing it to swell. Eventually the pressure builds until the ascus tip ruptures and the spores are ejected.

With some large-cup fungi, spore release can be a puff that is easily seen *and heard*, sometimes dramatically so. Once a fruit body hymenium is mature and asci are ready to fire, a simple disturbance of air may be all that is necessary to make all of them discharge simultaneously. Even the most ardent mycophobe will be pleasantly surprised to watch as you hold a carefully picked ascocarp fruiting body in front of you, blow a stream of air over its surface, and ... *puff!*

The initial release of spores can be quite active, even explosive!

MUSHROOMS WITH TEETH

Teeth may seem an unlikely feature for a mushroom, but they do exist. These mushrooms have a stalk and cap, but no gills. Instead, their spores are borne on teeth or spines that hang from the underside of the cap.

Hedgehog mushrooms (*Hydnum* species) are one example. Highly prized as delicious edibles, they are members of the chanterelle family, Cantharellaceae, with a similar overall shape and color when seen from above. Their spore-bearing spines or teeth stick out below the cap.

Hawk's wings (*Sarcodon* species) are beautiful but can be difficult to see in dark forests, as they blend in with their surroundings. They are stalked with gray teeth underneath.

A small, stalked tooth mushroom that you are unlikely to see but is worth the search is the earpick, *Auriscalpium vulgare.* Common but rarely encountered, it grows solely on the cones of pine and Douglas fir trees.

The northern tooth, *Climacodon septentrionalis,* is a stalkless toothed mushroom that grows on standing trees (mostly maples), with overlapping shelves often seen high above ground. The teeth are much tougher than other fleshy tooth fungi (*Hericium*). This mushroom can grow massively, often leaving a light area on the bark where the fruit body grew the previous year. It is not edible.

TOOTHED TREASURES

SPECIES LION'S MANE
Hericium erinaceus
SIZE Large cluster
HABITAT Live or dead wood
EDIBILITY Choice edible

SPECIES HEDGEHOG MUSHROOM
Hydnum repandum
SIZE Small to medium
HABITAT Soil near trees
EDIBILITY Choice edible

SPECIES HAWK'S WING
Sarcodon imbricatus
SIZE Small to medium
HABITAT Soil near trees
EDIBILITY Choice edible

Toothed mushrooms come in many forms, but all produce spores from spines that hang down, rather than gills or tubes. Many have a tough texture and are drably colored, though some are colorful. Many are saprotrophic, but some are biotrophic, such as the popular edible hedgehog mushroom.

HABITAT Conifer cones
EDIBILITY Not edible
SIZE Small
SPECIES EARPICK MUSHROOM
Auriscalpium vulgare

SPECIES CAT'S TONGUE
Pseudohydnum gelatinosum
SIZE Small
HABITAT Rotting wood
EDIBILITY Edible

SPECIES FUSED TOOTH
Phellodon confluens
SIZE Small to medium clusters
HABITAT Soil near trees
EDIBILITY Not edible

SPORE CATAPULTS

A drop of liquid sits above a water-sensitive trigger. Little by little, the drop grows larger, getting closer and closer to the trigger, until at last they touch, and the payload is flung out into the air. It sounds like some bizarre medieval weapon, but this is ballistospory, the process by which some basidiomycete fungi spread their spores.

If ascomycete fungi have squirt guns, basidiomycetes could be better described as a surface tension catapult. As the word "ballistospory" suggests, it's an explosive process. The spores, or ballistospores, develop on special tips called basidia. Atop each club-shaped basidium are outgrowths known as sterigma.

Key to the whole process is something called a Buller's drop. This is a small quantity of a sugary liquid, which grows on the sterigma. Moisture from the air condenses on this liquid, so the drop slowly gets bigger, held into a droplet shape by surface tension. At the same time, a film of water forms across the surface of the spore.

The liquid droplet—Buller's drop—grows until it reaches a critical size, at which point it touches the water film on the spore surface. At this moment, surface tension quickly pulls the drop onto the spore, the drop collapses, and the surface energy is converted to kinetic energy, pulling the spore from the hymenial surface and launching it into the air. Ballistospores are literally blasted from their basidia, but only for a short distance, before drag takes over and the spore decelerates. Released from the mushroom cap, it is carried away by air currents.

While the formation requires moisture from the air, too much water can disrupt this mechanism altogether. For this reason, many basidiomycetes have fruit bodies that are umbrella-shaped, to shield the hymenium from rain. Many others sequester the hymenium altogether within a fruit body, such as a puffball or a truffle. This is also why we typically do not find aquatic ballistosporic mushrooms.

PUFFBALLS AND TRUFFLES: Many fungi produce ball-like forms that protect their spore, hiding them on the inside like this *Calvatia* puffball (above). Truffles also sequester their spores inside and remain underground like the highly prized *Tuber* (right).

WIND, WATER, AND COW PIES

Fun though it is to witness, ballistospory isn't for everyone. Other basidiomycetes use different methods to get their spores out into the big wide world. From raindrops to ruminants, the only constant is variety.

Pilobolus species can fire their spores an amazing distance of 6 feet 7 inches.

PUFFBALL PROJECTORS: Although they don't have true stalks, earthstars are puffballs that can hoist their spore chambers above the forest floor as conditions become humid and conducive to spore release.

Gasteroid fungi make up the group of basidiomycetes that includes stinkhorns, puffballs, and bird's nest fungi. All of them require wind or water—or a well-placed kick—to release spores. Most puffballs puff out their spores through a hole in the top of the fruit body. Cute little bird's nest fungi produce their spores in small packets (peridioles) that appear as "eggs" in a nest-like cup. A raindrop hitting the cup will splash the eggs several inches or even a few feet.

Many other fungi release all their spores within a single tidy packet. Some coprophilous fungi produce spores within resistant packets that can pass through grazing animals and reemerge from the other end, along with a nice, fertile substrate in which to grow. (Yes, we are talking manure.)

But how does a fungus, growing in dung, get its spores into another ruminant? When it comes to grazing animals such as cattle, this is no easy trick, since cattle are careful to avoid one another's feces.

If you have been to a pasture, you may have noticed a "zone of repugnance" (the actual technical term) for the lush grass that goes ungrazed in the immediate vicinity of cow droppings.

Many coprophilous fungi get around this by firing their spore packets toward the light, sometimes over tremendous distances—8 inches in the case of *Podospora* species, 12 inches for *Ascobolus* species, and an amazing 6 feet 7 inches for *Pilobolus* species. The spores float away from the manure where the parent fungus grows, into fresh grass that cattle are happy to graze.

THE MUSHROOM MISSION

Every single mushroom has one mission:
to spread its spores. The amazing variety of shapes
and textures we can see are all defined by how
many spores they can grow, and how effectively
they can spread them around.

BRACKET FORMS:
Polypores often produce
durable, woody fruit bodies
that persist on their host
tree for many years. Spores
are borne on a fresh
hymenium annually.

CORALOID FORMS:
Many fungi resemble other
organisms such as plants
and animals. Some resemble
delicate corals, such as these
Clavulinopsis fusiformis.

Mushroom shapes mostly fit into a few common types. They are often listed as fruit bodies with gills, pores or tubes, teeth or spines (agarics and boletes); bracket and shelf-like mushrooms with pores or gills (polypores); bird's nest and cup fungi; puffballs and puffball-like fungi; jelly fungi; coral and club fungi; and truffles and truffle-like fungi.

In the past, fungi were often classified by their shapes, but we now know these can be misleading. Both ascomycete and basidiomycete groups include species that produce similar-looking mushrooms, such as brackets, cups, clubs, and truffles—a fascinating example of convergent evolution.

Polypore fungi live inside a host tree, and when it is time to reproduce, they produce a shelf-like mushroom from the side of the tree. Bracket and shelf fungi often have interesting physiologies and many are perennial, persisting on their woody hosts year round.

These fungi are commonly found growing from the main stems of living trees, so people often think of them as parasites or symptoms of disease. But more often than not, they are saprobes, feeding on parts of the tree that have already died. Many continue to grow on wood long after the host tree is dead, and they can grow to enormous sizes, often larger than a dinner plate.

Historically, some polypores have been used to create medicines, leather- and felt-like textiles, and as firestarters. *Fomes fomentarius* is one such polypore. Known as the tinder polypore, this fungus produces large brackets that are commonly seen throughout the Northern Hemisphere. The brackets have been used for fire starting since probably the Paleolithic era, 15,000 years ago.

PUFFBALLS PUT THE FUN IN FUNGI

Puffballs are widely familiar from lawns and sports fields. They are saprobic, feeding on plant matter such as dead grass.

Most are white and round, and some can grow to enormous sizes. The most curious thing about them is they have no visible gills. So where do the spores come from?

Puffballs have an advantage over many other types of mushroom-producing fungi, in that they shelter their hymenium (spore-producing surface) within a chamber, protected, as they dry out in the sunny open areas where they grow. When mature, the puffball skin (perideum) tears or breaks open to reveal the brown spore mass inside. When struck by raindrops, this mushroom puffs out clouds of spores that land nearby and begin the next generation.

While most puffballs are small, some are among the largest of all mushrooms. The surface of the giant puffball, *Calvatia gigantea*, is round, smooth, and white when fresh, becoming olive to brown when mature. Average specimens are about the size of a soccer ball (12 inches in diameter), but can be basketball-sized or larger. This species is found in woods, fields, or pastures in late summer and fall.

MEET THE PUFFBALLS

SPECIES GIANT PUFFBALL
Calvatia gigantea
SIZE Large to very large
HABITAT Open, grassy areas
EDIBILITY Edible and popular

SPECIES GEMMED PUFFBALL
Lycoperdon perlatum
SIZE Small to medium
HABITAT Woody debris and soil
EDIBILITY Edible

SPECIES PEAR-SHAPED PUFFBALL
Lycoperdon pyriforme
SIZE Small to medium
HABITAT Rotting wood
EDIBILITY Edible

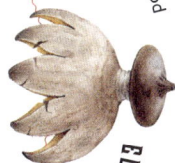

Puffballs and similar mushrooms have a simple, gasteroid (stomach-like) form, often white and saprotrophic, growing from debris; some are biotrophic. While not particularly tasty, most are edible, like giant puffballs. However, some, such as the toxic false puffball (or pigskin puffball), are poisonous. Caution is advised when identifying them.

SPECIES FALSE PUFFBALL
Scleroderma citrinum
SIZE Small to medium
HABITAT Mycorrhizal
EDIBILITY Toxic

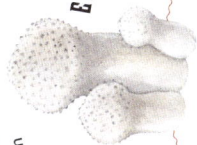

SPECIES EARTHSTAR
Geastrum triplex
SIZE Small to medium
HABITAT Woody debris and soil
EDIBILITY Not edible

SPECIES STALKED PUFFBALL
Tulostoma fimbriatum
SIZE Small to medium
HABITAT Woody debris and soil
EDIBILITY Not edible

TRUFFLES: THE MOST PRIZED FUNGI

Truffles look like no other mushrooms. They grow entirely underground and resemble small potatoes. The vast majority of truffle species are known from Australia, with more being discovered all the time. The famous and prized edible truffles of the world are mostly native to the Mediterranean and the West Coast of North America, but truffles are farmed in many other regions, including China and the Middle East.

Ripe truffles produce strong, savory odors that entice mammals to dig them up and consume them.

MUSHROOM IN DISGUISE: Truffles are mushrooms that are produced underground; indeed they closely resemble clods of dirt. But their odor is a giveaway. To see their hymenia, you need to slice a truffle in half. Inside, convoluted tissues—resembling mushroom gills in some cases—are where spores are born.

Truffles produce their spores underground, on a hymenium wrapped inside an external skin. So, when it is time to launch their spores into the air, they produce strong, savory odors that entice mammals to dig them up, consume them, and distribute their spores later (with the rest of their solid waste). Why switch to a life underground? Most truffles are found in drier regions, and producing a fruit body underground protects the spores from drying out.

Truffles are mycorrhizal fungi, partnering with specific trees or other plants. Truffle hunters have long known this and during the season search around tree species known to host truffles. To home in on the exact location of a ripe truffle, hunters historically used pigs but nowadays mostly use dogs.

FUNGAL KINGS AND QUEENS

Unlike truffles and puffballs, boletes look like "normal" mushrooms, with a cap on a stalk. Look underneath the cap, however, and you won't find any gills. Instead, boletes have a layer of tubes ending in a surface of pores from where spores are released into the air. Boletes make up an enormous group of macrofungi found in every forest.

One of the most prized edible mushrooms is the king bolete. It goes by dozens of names around the world and is actually a group of many similar species. *Boletus edulis* of Europe is the most widely known, but there are at least a dozen others that are nearly identical. All are large, growing to weights of 3 pounds and reaching diameters of 15 inches. They all have white pores when young. The pore surface is compact and looks cottony. As these mushrooms mature, the pores turn olive yellow-green, never pinkish. Near the top of the stem is fine reticulation—a net-like pattern of raised tissue. No boletes in this clade stain or bruise when cut.

The king boletes of North America range from almost white, *Boletus barrowsii*, to tan and brown, to reddish brown, *B. rubriceps* of the Rocky Mountain region. All feature a fat stem (often wider than the cap) with white reticulation near the top.

FUNGAL ROYALTY

SPECIES KING BOLETE
Boletus edulis
SIZE Large
PORE White pores become yellowish
STEM White reticulation

SPECIES ORANGE BIRCH BOLETE
Leccinum scabrum
SIZE Large
PORE Gray pores aging brown
STEM Rough scabers

SPECIES LARCH BOLETE
Suillus grevillei
SIZE Medium
PORE Yellow pores
STEM Partial veil

Boletes are stalked pored mushrooms and make up an enormous group of macrofungi; identification can be difficult. Pore colors (including white, yellow, or red), presence of a partial veil covering pores, stalk texture (smooth, rough, reticulated), and whether the mushroom stains blue or not are useful characteristics.

SPECIES BITTER BOLETE
Tylopilus felleus
SIZE Large
PORE White pores become pink
STEM Brown reticulation

SPECIES RAVENEL'S VEILED BOLETE
Pulveroboletus ravenelii
SIZE Medium
PORE Yellow pores
STEM Partial veil

SPECIES OLD MAN OF THE WOODS
Strobilomyces floccopus
SIZE Medium to large
PORE Black pores
STEM Partial veil

SCABER STALKS AND BITTER BOLETES

Not all boletes are as prized as the king—some are deeply unpleasant to eat, and others are toxic. Another group, "scaber stalks," is known for the rough protrusions, or "scabers," that grow on its stems.

A common species of scaber stalk, found across Europe and northern North America, is *Leccinum scabrum*. It is found exclusively with birch trees, hence its common name, brown birch bolete. The cap is dull brown when young but cracks with age, revealing the nearly white flesh beneath. The pore surface is pale gray and bruises brown when handled.

Other Leccinums occur with aspen trees or oaks. *Leccinum vulpinum* is a striking orange-red scaber stalk common with pines and spruce. Some *Leccinum* species have been known to make people sick, especially when eaten raw.

NO PORCINO:
Boletes like this *Tylopilus felleus* resemble king boletes, but these bitter boletes are definitely not palatable.

ROUGH STALKS:
The ubiquitous scaber stalks, *Leccinum* species, are easy to tell from other boletes— you need only feel the roughened stems.

Mushrooms in the genus *Tylopilus* are extremely bitter. They are often mistaken for good edible boletes, but one bite will change your mind. *Tylopilus felleus* is a common mushroom with a strong physical resemblance to members of the king bolete group but is definitely not as tasty. Looking at the reticulation on the upper stem, you will see that the highest edge of the pattern is darker than the background, exactly the opposite of the good edibles.

This mushroom is found with conifers and can occasionally be found growing from decayed stumps. The pore surface is white when young but turns pink with age and bruises dark pink when handled.

STICKY CAPS AND SATAN'S BOLETES

Boletes are an astonishingly varied group, some slimy or sticky, others elaborately decorated. All collectors should be wary of the aptly named Satan's boletes.

The slippery jack and its relatives are boletes in the genus *Suillus*. They can be found with conifers across the northern half of the globe. Most are small to medium, rarely exceeding 6 inches in diameter. They get their common name from having a slimy or sticky feel to the cap. There are many *Suillus* species, but the mushroom referred to as the slippery jack, *Suillus luteus*, is common under pine trees.

If you are not proficient with bolete identification, be careful of any that bruise blue. A good bet is to avoid blue-staining boletes with red pores. Deadly species are collectively known by a foreboding name: Satan's boletes. *Rubroboletus (Boletus) eastwoodiae* and *R. pulcherrimus*, in North America, have lightly colored caps and a striking red stalk and pores. The "true" Satan's bolete of Europe is *Rubroboletus (Boletus) satanas* and is similar, but the cap is white. All are among the prettiest of boletes, if not of all mushrooms—but looks can be deceiving.

Some boletes are wildly decorated with exaggerated features. Some have exaggerated stem reticulation, while others have elaborate tufts on the tops of their caps.

One famous ornamented bolete is the old man of the woods, *Strobilomyces floccosus*. They are gray with black decorations on the cap and a cottony partial veil that protects the developing gills. This often leaves a ragged edge on the mature cap. The flesh of some species of *Strobilomyces* stains an odd, reddish color when exposed to air. The red becomes darker until it appears black. These beautiful mushrooms can be found widely in the Northern Hemisphere.

A mushroom with exaggerated reticulation of the stem is Russell's bolete, *Aureoboletus* (*Boletellus*) *russellii*. The texture of the stem is garish, boasting ridges and valleys that are surprisingly large. This striking mushroom is abundant during some years and absent in others. It is most common in eastern North America but may be present in eastern Europe.

MAD ABOUT MORELS

Morels are the most popular of wild edible mushrooms. There are three groups: black, yellow, and half-free, with many species in each.

All but a few look identical. The difference is based not so much on color as on the shape of the convoluted cap: with yellow morels the bottom of the cap is attached directly to the stem; black morels have a loose overhang (vallecula) at the bottom of the cap; half-free morels have the bottom half of the cap hanging down, free from attachment with the stem.

Half-free morels could be confused with Verpas; a species with a smooth cap is *Verpa conica*, while *Verpa bohemica* has a convoluted cap. Both are close relatives of true morels but distinguished by having a hollow stem stuffed with cottony fuzz.

Be sure to note differences between true morels and false morels if you are collecting for dinner: although not deadly, some false morels are poisonous and can cause sickness. All true morels can be gray in color when immature and have a large hollow cavity and deeply pitted caps. False morels are not hollow and have a wavy appearance.

MYRIAD MORELS

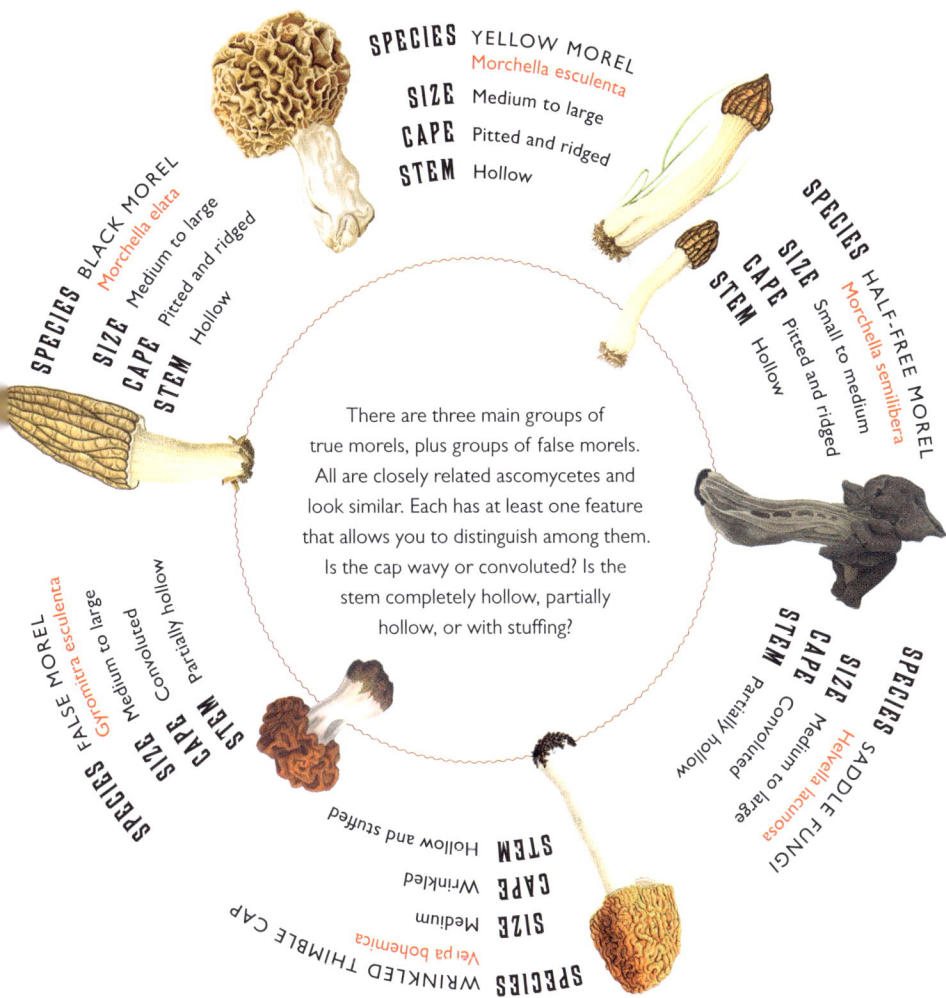

SPECIES YELLOW MOREL
Morchella esculenta
SIZE Medium to large
CAPE Pitted and ridged
STEM Hollow

SPECIES BLACK MOREL
Morchella elata
SIZE Medium to large
CAPE Pitted and ridged
STEM Hollow

SPECIES HALF-FREE MOREL
Morchella semilibera
SIZE Small to medium
CAPE Pitted and ridged
STEM Hollow

SPECIES SADDLE FUNGI
Helvella lacunosa
SIZE Medium to large
CAPE Convoluted
STEM Partially hollow

SPECIES WRINKLED THIMBLE CAP
Verpa bohemica
SIZE Medium
CAPE Wrinkled
STEM Hollow and stuffed

SPECIES FALSE MOREL
Gyromitra esculenta
SIZE Medium to large
CAPE Convoluted
STEM Partially hollow

There are three main groups of true morels, plus groups of false morels. All are closely related ascomycetes and look similar. Each has at least one feature that allows you to distinguish among them. Is the cap wavy or convoluted? Is the stem completely hollow, partially hollow, or with stuffing?

BEWARE OF IMITATIONS

False morels, or *Gyromitra*, are infamous as the wolves in sheep's clothing of the fungus world. Many look somewhat similar to true morels, or *Morchella*, and they are in fact closely related. However, most books list all false morels as toxic and warn against eating them.

In truth, some species seem to be safe edibles and are commonly collected for the table. Indeed, many people regard them as tastier than true morels! Based on scientific analysis, it seems that *Gyromitra infula* (which occurs on rotting wood) and *Gyromitra esculenta* may be the only species with toxins—hydrazine compounds similar to those in rocket fuel. These species should be avoided.

FALSE MORELS:
To distinguish false from true, false morels are not completely hollow inside, from top to bottom.

Some species contain hydrazine compounds, similar to rocket fuel.

All others, if eaten, should be fully cooked, as any traces of volatile hydrazines are driven off with heat.

The largest of the false morels is the red morel (also called big red), *Gyromitra caroliniana*, which may exceed 8 inches and have enormous stems. Big reds are found in mixed forests in eastern North America, as are pig's ears (also called gabled false morels), *Gyromitra brunnea*. The latter is a stocky false morel (about 5 inches) but more brownish, with the appearance of a tri-cornered hat. The conifer false morel, *Gyromitra esculenta*, is common, reddish to very dark with a slender stalk, and grows to about 4 inches.

Helvella species are also ascomycetes and are fairly closely related to *Morchella* and *Gyromitra*. Sometimes they also are called false morels but they are better known as saddle fungi or elfin saddles for the overall morphology of most species. *Helvella* mostly fruit in summer through fall and can be found in a variety of habitats including mixed forest as well as disturbed areas, along paths and grassy areas, and in compacted and even gravelly soils. Until recently, all Helvellas with a fluted stalk and gray to black convoluted cap were called *Helvella lacunosa*, a common species known from across North America and Europe. It has recently been determined that it probably does not occur in western North America. The fluted white helvella, *Helvella crispa*, is similar but with a white cap. There are many helvellas with smooth stalks. Probably the most common is *Helvella elastica*, which grows to about 4 inches.

MEET THE RELATIVES

We may look nothing alike, but fungi and animals share many similarities.

Across all of the five kingdoms of life, fungi and animals are the closest relatives. We have more in common than you might expect!

For a start, we share a common ancestor. Although fungi and animals look nothing alike in body shape, nor at the cellular level, chemically and physiologically, we share many similarities. Unlike animals, fungal cells have cell walls; but inside the cell, the chemistry is similar—there are ribosomes, mitochondria, and DNA is organized as chromosomes. Like animals, fungi produce an assortment of enzymes and can digest and catabolize most substances that we can. Unlike animals, fungi produce cellulase enzymes and can digest wood.

Fungi perform a valuable service by breaking down and recycling all that cellulose from dead trees and other plants. But this was not always the case—as demonstrated by the fact we can dig up vast pockets of fossil fuels. Fungi capable of rotting woody plants did not appear on the scene until the end of the Carboniferous Period (359–299 million years ago)—quite a bit after the evolution of woody plants. So instead of decomposing, all that organic matter became fossil fuels such as coal.

IT'S A ROTTEN WORLD OUT THERE . . . THANKS TO FUNGI

We can't eat wood. Or at least, we can't extract much nutrition or energy from it. Plants are mostly made up of tough materials called cellulose and lignin. Both are difficult to break down and require arsenals of enzymes and other machinery—arsenals developed by fungi.

For the most part, wood-degrading fungi are good at breaking down either cellulose or lignin. Lignin is a polymer of tough ring molecules crosslinked in a random fashion that strengthen wood. Fungi that directly break down cellulose, leaving the brown lignin behind, are called "brown rot" fungi. The removal of the cellulose destroys the structural integrity of the wood and it cracks and falls apart. Examples found worldwide are the polypores *Laetiporus* (sulfur shelf), *Phaeolus schweinitzii* (dyer's polypore), and *Fomitopsis* (belted polypore).

Tough wood fibers are no match for fungi's arsenal of digestive enzymes.

BEAUTIFULLY ROTTEN:
The sulfur shelf (_Laetiporus sulphureus_) is a colorful polypore mushroom found growing on decaying wood worldwide.

WORLDWIDE ROT:
Numerous white-rot fungi,
such as these turkey tails, *Trametes
versicolor*, are commonly seen in
just about every forest on Earth.

The other variety are the "white rot" fungi that decompose lignin, bleaching the wood and initially leaving stringy white cellulose behind. These fungi have powerful peroxidase and laccase enzymes that break down lignin. While there is evidence that they can decompose it completely, many researchers suggest the fungi are mostly removing lignin from the woody pulp to improve access to the cellulose.

White-rot fungi found worldwide include polypores such as *Inonotus* (chaga and others), *Ganoderma* (artist's conk), and *Trametes* (turkey tail), as well as *Pleurotus* (oyster mushroom) and *Armillaria* (honey mushroom), as is the popular cultivated shiitake, *Lentinula edodes*. The power to bleach wood pulp of lignin makes the white-rot fungus *Phanerochaete chrysosporium* important to the paper industry as an environmentally benign replacement for harsh synthetic chemicals.

Many of the wood-rot fungi familiar to us (for example, the big polypores) don't wait for trees to die before launching their assault. Living trees often sport big fruit bodies of shelf fungi. This is because most of the tree is heartwood—the dead inner wood. (Only the outer layers, just under the bark, are living tissue.) All it takes is a wound to disrupt the integrity of the bark, and heart rot can ensue.

Walking through the woods, you may have noticed the results of both white-rot and brown-rot fungi. It's likely you thought the stringy white pulp and the cuboidal-blocky chunks left behind were due entirely to the species of tree being decomposed. That is part of the story, but the main reason is actually the fungi that were involved.

SHEDDING LIGHT ON BIOLUMINESCENCE

Wooden beams in a coal mine glowed so brightly that pit lamps were unnecessary.

Glowing mushrooms are amazing to see, even for mushroom haters. The phenomenon is known as bioluminescence, and it has been documented since ancient times.

Although bioluminescence was mentioned by Aristotle and Pliny the Elder, naturalists mostly ignored the subject until the eighteenth century. In 1796, Alexander von Humboldt was intrigued by tales of miners in Germany. Wooden panels and beams in a coal mine were reported to glow—so brightly, in fact, that pit lamps were unnecessary. The glow occurred when humidity and temperatures within the mineshaft were high. Light was described as being emitted by "plants" (termed *Rhizomorpha* species).

We now know that the source of the glow was fungal. There are four known groups of bioluminescent fungi, containing around eighty different species altogether. Mushrooms that glow include *Armillaria*, *Mycena*, *Omphalotus*, and *Panellus*. If the light comes from hyphae in wood, often called "foxfire," it is most likely species of *Armillaria*.

Bioluminescence is widespread in nature; in addition to fungi, some animals, plants, and bacteria do it. Two things to keep in mind about bioluminescence are that it is ongoing, even in the light of day, when it is not visible; and it generates no heat, and so is different to incandescence, which is a thermal glow. Both Conrad Gessner (Swiss physician and naturalist, 1516–1565) and Francis Bacon (English philosopher and statesman, 1561–1626) recognized that the light emission from wood was not related to heat radiation.

Today it is known that the light originates from a metabolic reaction of the fungus where electrons are transferred to an acceptor molecule (luciferin), which is cleaved by an enzyme (luciferase) in the presence of oxygen. This results in the formation of an electronically excited state of the luciferin and the subsequent emission of light with a maximum wavelength of approximately 525 nanometers during return to the ground state. This process is much the same for all organisms that bioluminesce, though the luciferins and luciferases are not exactly the same.

WHY DO MUSHROOMS GLOW IN THE DARK?

A number of theories have been put forward for why some fungi glow. A number of them suggest it is to attract invertebrates to spread the spores. There's some evidence this might be true of species in thick tropical forests, where there is little wind to spread spores, but it doesn't seem to be the case in temperate environments.

Bioluminescence may simply be a way for fungi to dissipate energy generated by their metabolism (as most organisms, including us, give off heat as a by-product). Or the reaction may be linked to neutralizing toxic peroxides, which are formed as wood decays. Many bioluminescent fungi, such as *Armillaria mellea* and *Panellus stipticus*, are white-rot fungi (see page 81), and scientists have found that limiting their ability to process lignin also reduces their bioluminescence.

Despite all this, the function of bioluminescence remains elusive and controversial. Some researchers suggest there is no evolutionary benefit resulting from bioluminescence in fungi, as there are groups with glowing and nonglowing species that all seem to be equally successful in nature. In that case, it seems likely that bioluminescence was an

evolutionary benefit within a group of fungi in the distant past—maybe for spore dispersal—and has been retained as evolutionary "baggage."

Within the genus *Mycena* there are at least thirty-three species, from sixteen sections, known to bioluminesce. Many more do not, of course. And this begs the question: did bioluminescence evolve once and was a trait lost many different times throughout history, or did it evolve many different and independent times within the genus? My guess is that the trait seems to be of some benefit as it is retained in so many species of fungi—but your guess is as good as mine.

GLOWING MYCENAS:
The genus *Mycena* is a large group of small, mostly saprobic mushrooms found in dense forests, including tropical rainforests. There are at least thirty-three species, from sixteen sections, known to glow, making *Mycena* the most bioluminescent group of fungi. Undoubtedly, many more await discovery.

DEATH BY MUSHROOM

Mushrooms serve as a significant (and tasty) food source across many cultures and countries. They are also known to be killers, bringing swift death if the wrong variety is eaten.

While murder-by-mushroom was a staple of ancient assassins, these days accidental poisoning is more common, as distinguishing between edible and poisonous species can be challenging. Additionally, some people may have an allergy or sensitivity to wild mushrooms (as can happen with any foodstuff), and there are a few enigmatic mushrooms that normally are well-known as safe edibles but when combined with other foods such as alcohol or dairy can become mildly toxic.

Mushroom poisoning is a health concern worldwide, leading to both morbidity and mortality, and despite improved education, the numbers of poisonings continue to rise—no doubt as a result of the ever-increasing popularity of foraging, mushrooms especially. While the global fatality rate from mushroom consumption is unknown, it is speculated to be at least a hundred deaths per year. This likely underestimates the actual impact, given the approximate fifty to a hundred deaths annually in Europe alone. Notably, amatoxin poisoning—a type of mushroom toxicity causing hepatocellular (liver) damage—poses a significant burden on healthcare systems

TOXIC TOADSTOOLS

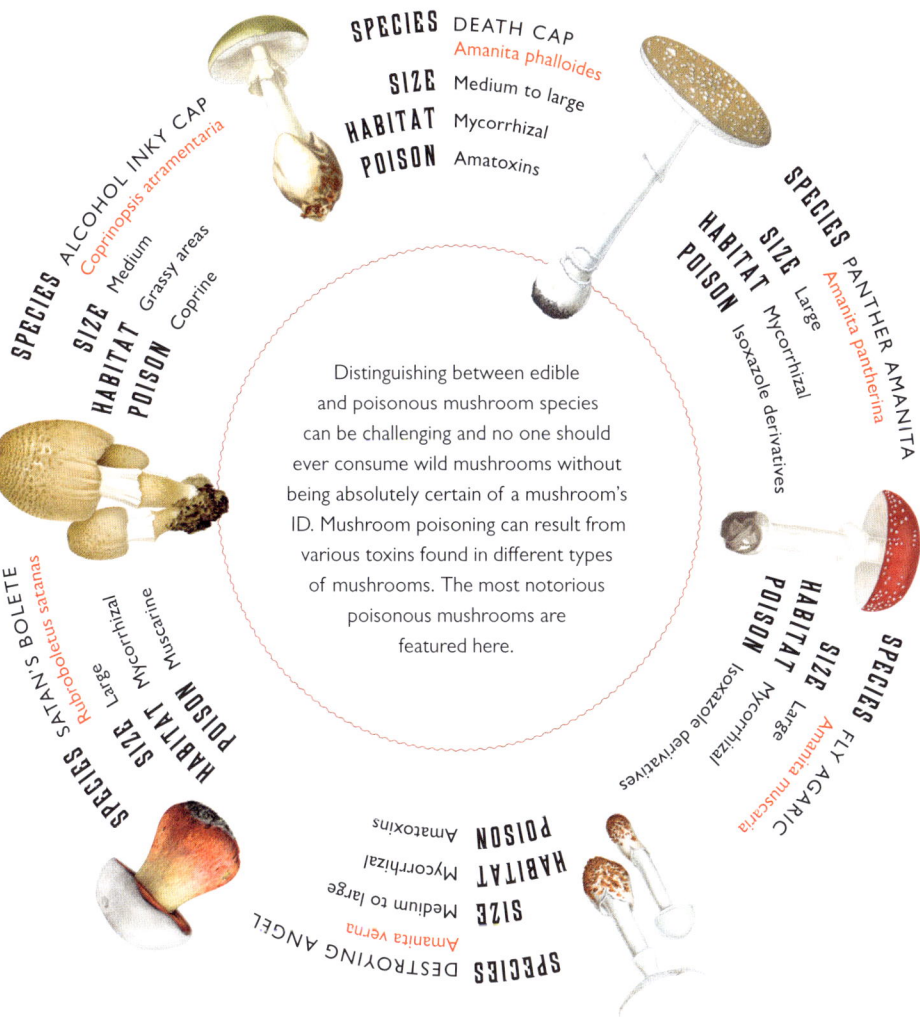

SPECIES DEATH CAP
Amanita phalloides
SIZE Medium to large
HABITAT Mycorrhizal
POISON Amatoxins

SPECIES ALCOHOL INKY CAP
Coprinopsis atramentaria
SIZE Medium
HABITAT Grassy areas
POISON Coprine

SPECIES PANTHER AMANITA
Amanita pantherina
SIZE Large
HABITAT Mycorrhizal
POISON Isoxazole derivatives

SPECIES FLY AGARIC
Amanita muscaria
SIZE Large
HABITAT Mycorrhizal
POISON Isoxazole derivatives

SPECIES SATAN'S BOLETE
Rubroboletus satanas
SIZE Large
HABITAT Mycorrhizal
POISON Muscarine

SPECIES DESTROYING ANGEL
Amanita verna
SIZE Medium to large
HABITAT Mycorrhizal
POISON Amatoxins

Distinguishing between edible and poisonous mushroom species can be challenging and no one should ever consume wild mushrooms without being absolutely certain of a mushroom's ID. Mushroom poisoning can result from various toxins found in different types of mushrooms. The most notorious poisonous mushrooms are featured here.

While murder-by-mushroom was a staple of ancient assassins, these days accidental poisoning is more common.

UBIQUITOUS FLY AGARIC:
Popular in science and pop
culture, the fly agaric is incredibly
common and widespread,
known on all continents
except Antarctica.

and contributes to liver transplantation cases. Although global data is lacking, local studies do suggest an increasing incidence of mushroom poisoning.

An emerging risk relates to large migrant populations foraging for food due to economic hardship—as well as confusing foreign mushrooms with look-alikes from "back home." Consumption of unfamiliar mushrooms by these migrants has led to cases of poisoning, particularly amatoxin-type mushroom poisoning involving species of *Amanita* (death caps and their ilk).

There are several other mushroom poisoning syndromes that are infrequently seen and poorly understood including: rhabdomyolysis or hypoglycemic poisoning resulting from ingestion of certain historically well-known and popular edibles such as man on horseback (*Tricholoma equestre*) and *Trogia venenata*, respectively; trichothecene poisoning involving mere handling of poison fire (*Trichoderma cornu-damae*); paxillus syndrome from ingestion of poison pax (*Paxillus involutus*); encephalopathy syndrome from ingestion of angel wings (*Pleurocybella porrigens*); and even mushroom dermatitis from handling shiitake (*Lentinula edodes*)—one of the most-consumed mushrooms on the planet—and certain other mushrooms.

TYPES OF MUSHROOM POISONING

Poisoning can result from various toxins found in different types of mushrooms. There are eight major syndromes.

Gastrointestinal: The most common form of mushroom poisoning occurs due to a wide range of gastrointestinal irritants. Symptoms include nausea, vomiting, cramps, and diarrhea, typically appearing within 20 minutes to 4 hours after ingestion. Not life threatening.

Muscarine: Certain mushrooms, including *Inocybe* and *Clitocybe* species, have been found to produce muscarine in intoxicating levels. Trace amounts also are produced by *Rubroboletus* (*Boletus*) *satanas* and other red-pored boletes, and *Amanita* species including *Amanita muscaria* and *A. pantherina*. Ingestion of muscarine can lead to excessive salivation, sweating, teary eyes, severe vomiting, diarrhea, and in rare cases death.

Isoxazole derivatives (muscimol, ibotenic acid, and relatives): These compounds cause hallucinogenic effects and are found in mushrooms such as *Amanita muscaria* (fly agaric) and *A. pantherina* (panther amanita). Symptoms include altered perception, ataxia, nausea, confusion and delirium, vivid dreams, and altered consciousness.

Amanitin (amatoxins): *Amanita phalloides* (death cap) and related species (destroying angels) contain amatoxins, which cause severe liver and kidney damage. Symptoms may not appear

DEADLY TRIO:
Amanita pantherina,
the panther amanita (right);
Amanita phalloides,
the death cap (below);
Gyromitra esculenta,
the false morel (bottom).

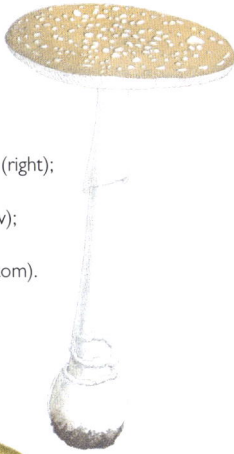

until 6 to 24 hours after ingestion; often fatal. These mushrooms are responsible for 90 to 95 percent of mushroom fatalities annually.

Gyromitrin: Some species (false morels) contain gyromitrin, which can lead to nausea, vomiting, abdominal pain, and seizures. Not life threatening.

Orellanine: Some species of *Cortinarius* (web caps) can cause kidney damage. Symptoms may appear 24 hours to 21 days after ingestion. Can be fatal.

Allenic norleucine: These toxins cause rapid kidney failure. Some species of *Amanita*, including *Amanita smithiana*, produce these compounds.

Coprine and other alcohol-induced syndromes: Consuming alcohol with certain mushrooms (e.g., *Coprinopsis atramentaria*, alcohol inky) can lead to disulfiram-like reactions, including flushing, nausea, and vomiting. Not life threatening.

THE "MAGIC" OF MUSHROOMS

While ancient mushroom gatherers were looking for food, and wary of poisons, there is of course a third use that humans have found for fungi. Certain mushrooms, known as entheogenic (or "magic") mushrooms, are infamous for causing hallucinations and altered perception—and they have become a hot topic of late.

These are naturally occurring mushrooms (*Psilocybe* species, and other genera including *Pluteus* and *Omphalotus*) that contain the chemical compounds psilocybin, psilocin, and other indole derivatives. These compounds are often listed as toxins in mushroom books. They are not. They are more correctly considered intoxicants or hallucinogens. For this reason and others, many are purposely consumed.

Furthermore, psychedelic mushrooms, and the powerful molecules they produce, were once and are now again the subject of intense study for their potential to cure a number of mental pathologies.

Roland Griffiths and Charles Grob of Johns Hopkins University wrote in the December 2010 issue of *Scientific American* ("Hallucinogens as Medicine") that, "In a matter of hours, mind-altering substances may induce profound psychological realignments that can take decades to achieve on a therapist's couch."

The American mycologist Gary Lincoff (1942–2018) had much to say about mushrooms of the genus *Psilocybe*. Upon being asked what he thought the future held for psilocybin as a therapeutic, Gary had fervent optimism. When asked to put into words the psychedelic experience, the overall feeling of love, and the connectedness with all life—the ineffable—in true Gary fashion, he always had much to say:

> *So, who in their right mind, you might ask, would take psilocybin? Someone out of their (left) mind? Or, if you are finding yourself on planet Earth in the twenty-first century, and are wondering who took the wrong turn, it's too late to check your genome. In this case, it might just be better to sit back, relax, take 5 grams of mushrooms and call me when the moon is in the seventh house and Jupiter aligns with Mars.*

Gary Lincoff,
"Psilocybin: Its Use and Meaning," *FUNGI Magazine*, 2011.

DECEIVING LOOKS:
Although *Psilocybe cubensis* looks like many other ordinary mushrooms, this species—the most commonly cultivated "magic mushroom"—has extraordinary chemical properties.

CHEMICAL MIND CONTROL: ZOMBIE FUNGI

One bizarre group of fungi merits a special note. These astonishing invaders have the ability, not only to infect insects, but to take control of their brains and alter their behavior. They are ascomycetes of the families Ophiocordycipitaceae and Clavicipitaceae, often dubbed "zombie fungi."

One of the most talked about species is *Ophiocordyceps sinensis*, an insect pathogen. These sinister fungi infect an insect host, grow as hyphae throughout the animal's body, and just before death, take control of the host's brain, instructing it how and where to move (the "zombification" part).

Ants parasitized by the fungus *Pandora formicae* move away from the colony. In most instances, the host is compelled to crawl up and latch on to a suitable growing point for the fungus (an act known as "summit disease"). A macabre death follows as the fungal sporophores erupt out of the host to launch their spores. Ants parasitized with *Ophiocordyceps unilateralis* position themselves on foliage directly above paths frequented by other members of their colony—infectious spores then rain down upon the unwitting next victims, below.

HIMALAYAN ZOMBIE: One bizarre group of pathogenic fungi, the "zombie fungi" (*Cordyceps* species), specializes in infecting insects and other arthropods. The most famous attacks the ghost moth caterpillars of the Himalayas, shown here.

Another zombifying fungus, of sorts, is *Massospora cicadina*. This strange entomophthoralean fungus was named by Charles Horton Peck, New York State Botanist from 1867 to 1915. He made 36,000 collections of fungi, mosses, ferns, and seed plants during his famed career, but *Massospora cicadina* is undeniably one of the strangest.

It infects periodical cicadas and produces psychedelic compounds that cause the host to fly around frenetically, attempting to mate with others of its kind—thereby spreading the infectious spores. As the zombified cicada host nears death, the fungal pathogen produces thick-walled resting spores . . . and then things get really gruesome. The terminal segments of the insect's abdomen fall off, and resting spores spill out. The spores will rest there in the soil for years—or decades—awaiting a new cicada juvenile to crawl through the soil.

RINGED MUSHROOMS:
Parasol mushrooms
(*Macrolepiota procera*) are also
commonly seen in open grassy
areas as well as forests, and
when mature can be identified
by their ring, which is loose on
the stalk and can be moved up
and down.

STARTLING STINKHORNS

Surely the most controversial group of mushrooms, stinkhorns are foul-smelling and slimy, they attract flies, and they grow to a shape that scandalized Victorian society. They are definitely mushrooms with character.

All stinkhorns are saprobic, feeding on rotting debris. They can be found in parks and urban areas, farms and pastures, and woodland areas as well. The smell comes from a mass of spores on the cap, called a gleba. Flies land on the gleba to dine, picking up spores on their legs, and depositing them on rotting matter elsewhere.

Many species of stinkhorns are placed in the genus *Phallus* for obvious reasons. Mycologists are forever getting phone calls and emails from people about strange phallic-shaped

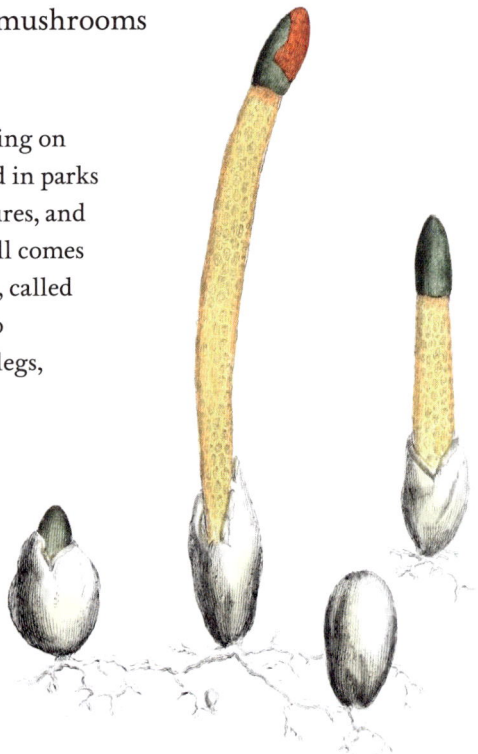

manifestations in their garden or flower beds. Shocking to some in appearance, they are beneficial in breaking down debris to create healthy soil.

A moral challenge

Throughout history, stinkhorns have been known to cause a stir—and not just because of their odor. This was especially so during Victorian times. The more puritanical members of society were often alarmed by these most prurient members of the fungal kingdom. Most famously, Charles Darwin's sister, Henrietta "Etty" Darwin, had a strong reaction to stinkhorn mushrooms. She would reportedly, stomp, bury, and even burn them to protect the morals of the maids and household staff.

STRONG-SMELLING FUNGI

Fungi can produce an astonishing range of smells, from aniseed to sewer gas. The smells can be pleasant or deeply offputting, but they can also be useful in identifying mushrooms in the wild.

While the vast majority of mushrooms simply have a mild "mushroomy" odor, some may be more pungent or even putrid. *Phyllotopsis nidulans*, the pretty orange mock oyster, has a sour smell, and *Tricholoma sulfureum* (stinky trich), has a repulsive sewer gas or coal tar smell. (Even so, it's hard to resist taking a second and a third whiff of those surprising smells!) *Amanita caesarea* (Caesar's mushroom) is a prized edible when young and fresh but can take on a fishy smell with age. Indeed, several mushrooms can have a fishy odor, including *Russula xerampelina* (shrimp russula) and several milk mushrooms.

Other mushroom species simply smell divine— indeed, some mushrooms have the most wonderful and alluring fragrances in nature. *Clitocybe odora* is a common mushroom of North America and Europe, and famously fragrant, smelling strongly of anise. So does *Hydnellum suaveolens*. *Tricholoma saponaceum* (soapy trich) is named after its soapy smell. Some smell fruity; the species name *Inocybe pyriodora* literally means "smells like pears."

ALLURING OR ACRID?

Mushroom odors are important characteristics used in their identification. Many have a mild "mushroomy" odor, some may smell sweet, while others may be foul. Some mushrooms smell like other foods, including garlic, cucumber, radish, potatoes, or even seafood. Note: the smell most often has no hint to edibility or palatability.

SPECIES MOCK OYSTER
Phyllotopsis nidulans
SIZE Small to medium clusters
HABITAT Rotting wood
ODOR Very sour

SPECIES CAESAR'S MUSHROOM
Amanita caesarea
SIZE Medium to large
HABITAT Mycorrhizal
ODOR Fishy with age

SPECIES ANISEED MUSHROOM
Clitocybe odora
SIZE Medium to large
HABITAT On soil near trees
ODOR Anise

SPECIES SHAGGY SCALYCAP
Pholiota squarrosa
SIZE Large clusters
HABITAT Rotting wood
ODOR Strong garlic

SPECIES FRAGRANT INOCYBE
Inocybe pyriodora
SIZE Small to medium
HABITAT On soil near trees
ODOR Pears

SPECIES STINKY TRICH
Tricholoma sulfureum
SIZE Medium
HABITAT Mycorrhizal
ODOR Foul sewer gas

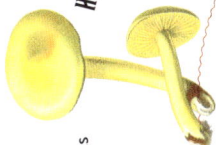

THE NOSE KNOWS:
Species of *Agaricus* are mostly
indistinguishable and typically
brown or white. But their odors
can vary tremendously from
sickly chloroform or phenol to
pleasantly mushroomy or sweet,
like almonds.

Several amanitas smell of raw potatoes, and wild mushrooms can have odors reminiscent of several other vegetables. *Macrocystidia cucumis*, a small brown mushroom, is easily keyed out given its smell of cucumbers. *Marasmius alliaceus* and *Pholiota squarrosa* smell strongly of garlic. The fiberheads (*Inocybe* species) lack a lot of showy characteristics but often make up for it with odors of green corn or fruity smells (though many others have a sickly spermatic smell). Many mushrooms have a raphanoid, or radish smell, including the poison pie, *Hebeloma crustuliniforme*. *Tricholoma apium* smells of celery.

The specific chemicals that give mushrooms their characteristic odors are best known in the truffles. The chemical 2,4-dithiapentane is the most abundant odorous compound in *Tuber magnatum*, the famous white truffle of Italy, and this is also the ingredient in commercial truffle oil. For those who have never been fortunate enough to have smelled culinary truffles, the odor is heady, musky, and complex with notes of Romano cheese thrown in.

A distinct odor such as phenol, iodine, or library paste warns that your white meadow mushrooms are of the *Agaricus xanthodermus* group, and not the more desirable *Agaricus campestris* group. Some amanitas in the *Lepidella* group smell strongly of chlorine bleach. *Cortinarius camphoratus* takes its name for a camphor-like smell.

Some species only emit an odor when dry, and the best example of this is the *Lactarius* species. *Lactarius fragilis* and *L. rubidus*, commonly called candy caps, are described as odorless when fresh; however, when heated or dried out, they smell aromatic, often described as the scent of maple syrup, burned sugar, fenugreek, or butterscotch.

This odor is also noted in some other species of *Lactarius*, also called milk mushrooms. One small, sordid (dull, dirty, or muddy colored), and otherwise nondescript species of milk mushroom is easily identified from its smell alone: *Lactarius glyciosmus* exudes a wonderful coconut odor.

The enticing smell of cherry or bitter almond comes from benzaldehyde and benzyl alcohols produced by several plants but also some mushrooms, including the big laughing gym, *Gymnopilus junonius*, and *Agaricus augustus*, the prince. The latter is a prized edible; the former is extremely bitter and unpalatable. If your mushroom smells strongly of maraschino cherries, then it's likely a brittle gill, *Russula fragrantissima*, or close relative. (They definitely do not taste anything like they smell!)

The best and the worst

The best-smelling mushroom? Arguably, it's a curiously small, white, and otherwise nondescript one called *Cuphophyllus russocoriaceus*, the cedar waxcap. The strong smell is reminiscent of cedar plus something else; many liken it to the smell of old leatherbound books. It takes its name from "Russian leather." Even after being dried for several years, specimens still give off a heavenly aroma.

And the worst? While a beautiful little mushroom in garish shades of almost fluorescent yellow-green, and with pink gills, *Entoloma incanum* probably holds the title of most foul. This mushroom emits the odor of mouse urine. Ever spent time in an old hayloft or dusty farmhouse attic? Then you know the smell.

Small, muddy, nondescript Lactarius glyciosmus *smells wonderfully of coconut.*

DECEIVING ODORS:
The little coconut-scented milk cap, *Lactarius glyciosmus*, is wonderfully fragrant but tastes nothing like it smells. This is the case with many odiferous mushrooms.

FRIENDSHIPS WITH FUNGI

We think of farming as a human activity, but some insects are known to grow fungus for food.

Fungi can poison, enslave, disease, and even enrapture animals. But they can also form amazing partnerships with them. Fungi and insects, in particular, have evolved many pairings where each benefits the other.

The scientific term for such partnerships is "coevolution." The great American evolutionary ecologist Daniel Janzen considers coevolution "an evolutionary change in a trait or traits of one organism, as a response to traits of another, different, species of organism."

What might have begun as parasitism or predation evolves over time into a much more benign relationship. In doing less harm to the host, for example, or even promoting its growth, the parasite benefits itself by ensuring a stable, long-term home. Endophytic and mycorrhizal fungi seem to be headed this way with their plant hosts.

There are countless situations where insects happen to transport fungus spores to a new growing site. However, most of the time, this is not really a display of coevolution, which is much less of a random act. Fungus-farming insects are a good example of coevolution in action.

We may think of farming as a human activity, but some ants, termites, and ambrosia beetles are known to grow fungus for food, and more examples are being discovered.

Some of the defining features of agriculture include planting (tilling or ploughing, and seeding, or "inoculation"), cultivation ("weeding" and removal of pests and disease), harvesting, and nutritional dependency. Amazingly, insect farmers exhibit these characteristics too! Their farming strategies include evolved mechanisms for preparing the ground, inoculation with crop spores, optimization of fungal growth through regular activities, protection of the crop against parasites or diseases, harvest, and consumption of fungi. Human agriculture, especially in a commercial setting, features a division of labor between cultivation, planting, and harvesting. Fascinatingly, the insect farmers have developed task-partitioned societies cooperating in their fungiculture biology. In ant and termite farmers, agriculture tasks are partitioned between different castes, each specialized to one main task. The result? A reliable supply of fungus food.

ANISEED MUSHROOM:
In addition to its blue-green color, *Clitocybe odora* has a wonderful odor of anise.

CHANTERELLES:
Cantharellus species are prized
edible mushrooms and many
species have an odor
of apricots.

MORE ABOUT FUNGAL FARMERS

If you have ever seen an ant or termite nest devoted to fungus farming, you will know what an astonishing sight it is. The processes involved are so delicate and specialized, it seems astonishing that they can be carried out without human intelligence. And yet, these partnerships date back millions of years.

Several species of New World ants in North America collect plant materials and use them to cultivate mushrooms in the genus *Leucoagaricus*. These insect farmers cultivate fungi in subterranean gardens, using the process of decomposition rather than photosynthesis to produce the nutrients that they need to survive. DNA analysis of seven ant species and their fungus partners suggests that the ants started

ANCIENT AGRICULTURE:
Termites have been cultivating *Termitomyces* fungi for tens of millions of years.

farming 55–60 million years ago. This long process of coevolution led the ants and fungi to become irreversibly dependent on the other; the ants lost their ability to produce the amino acid arginine on their own, and the fungi lost their ability to digest wood or bark and, therefore, rely on leafy plant matter delivered to them by the ants.

As an incredible result of convergent evolution, Old World termites cultivate fungal gardens in a similar manner to the New World ants. The termites cultivate *Termitomyces*, a genus that belongs to the family Agaricaceae. *Termitomyces* produce large to extremely large mushrooms and in some parts of the world are collected by humans as food.

Termites benefit in two ways from cultivating this fungus: one is that the fungus serves as food directly; the other is that it breaks down wood, and in particular cellulose, making food for the termites. Termites cannot digest wood in its natural form and depend on other organisms such as fungi to do so.

How do mushroom gardens get started? Most termites start their gardens from scratch when they begin a new colony, collecting spores from fruit bodies to begin their new crop. A few termite species take fungus from their home colony with them when they set out to start a new life.

Although termites resemble ants morphologically, they arose far earlier than ants and started farming fungus 30–50 million years before ants did. These termites evolved in Africa before spreading to Asia. We do not have such termites in the New World; all New World termites rely on fungi and other microbes in their gut to digest cellulose.

PARTNERS FOR PLANTS

Partnerships between fungi and animals are fascinating but relatively rare. We might expect the same to be true of plants, but the science increasingly shows that the plant and fungus kingdoms are deeply intertwined, forming a relationship that is vital to life on Earth.

The vast majority of plant species form a mutually beneficial living relationship with fungi. In fact, fungi—not roots—are the main channel by which nutrients travel from the soil into plants. It is likely that the roots of 90 percent or more of the world's plant species (and pretty much all trees) are colonized by symbiotic fungi.

Mycorrhizal (literally, "fungus root") associations involve fungal hyphae that grow from inside and around the roots of the host plant, outward into the surrounding soil.

Most plant species could not survive without their fungal partners.

This effectively increases the surface area of the root system by hundreds or even thousands of times. Mycorrhizae are so common and fundamental to plant nutrition that most plant species could not survive without their fungal partners unless artificially supported. (Conversely, in situations where abundant water and fertilizers are added, the plant may cast off its fungal partners, which is possibly why mushroom diversity is so much lower amid trees in urban settings.)

Essentially benevolent parasites, mycorrhizal fungi absorb food from the plant, then reward the plant for its hospitality by supplying water as well as essential nutrients such as nitrogen, phosphate, and potassium. Some mycorrhizal fungi have cellulase enzymes, and therefore probably gain nutrition saprobically from decay of organic matter in the environment, as well as biotrophically from their hosts.

LONG-DISTANCE PARTNERS: Although not always seen, the mycorrhizal fungal partners of plants send their hyphae through the soil at great distances from their host to scrounge water and nutrients.

AT THE ROOT OF THE MYSTERY

The hidden partnership between plants and fungi has shaped the world we live in, from the evolution of aquatic plants into land-dwellers, to the growth of forests and grasslands. This partnership dominates life on Earth to this day.

Fossils show that partnerships between plants and fungi date back around 460 million years, to the earliest days of land-based plants. In fact, the invasion of land-based habitats by plants couldn't have happened without fungi.

Aquatic plants were unable to survive the harsh conditions on dry land without help. The earliest land plants had no true roots, but they were colonized by hyphal fungi that formed

FOREST PARTNERS:
Many of our most prized edible mushrooms form ectomycorrhizal symbioses with trees.

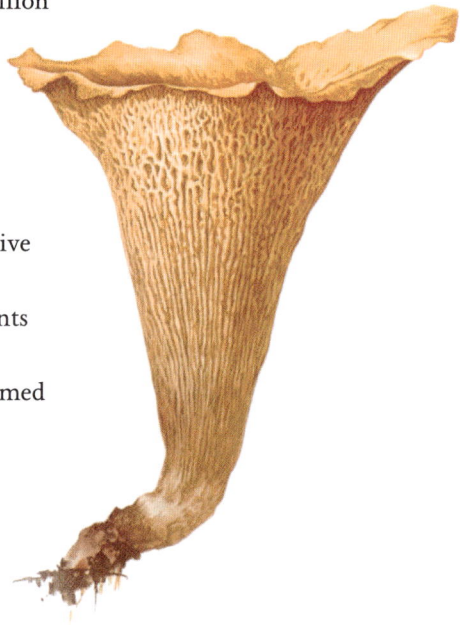

structures strikingly similar to modern mycorrhizas. From these lowly beginnings, terrestrial plants and mycorrhizal fungi spread together across the globe. Indeed mycorrhizal associations have evolved several times through the ages.

By definition, all mycorrhizas involve plant roots, but the physiology can be quite different across the spectrum. The two main groupings are known as ectomycorrhizal (EcM) and endomycorrhizal (EdM) fungi.

Ectomycorrhizal fungi grow into the plant root tissues but do not enter the root cells. Instead, the fungal hyphae grow around the outer cortical cells of the root, forming what is called a "Hartig net." Ectomycorrhizas exist most often as a mantle or covering of interwoven fungal hyphae on the surface of the fine roots of trees. The mantle makes the root tips look swollen and can be visible to the unaided eye. Today, EcM fungi are associated with most conifers and many hardwoods, including oaks, beeches, *Nothofagus*, and *Eucalyptus*. Well over four thousand species of EcM fungi occur in our forests across the globe, including many of our most prized edible fungi such as boletes, chanterelles, amanitas, and truffles.

Endomycorrhizal fungi grow into the plant root tissues and do penetrate the plant root cells. They do not produce a thick mantle over the surface of the root like EcM fungi, nor do they produce large showy fruit bodies. Indeed, most produce no real fruit body at all; a few EdM species produce balls or clumps of spores in the soil, while many seemingly do not undergo sexual reproduction, and may not even have the genes for it. Most EdM fungi are poorly known,

CHOOSY PARTNERS: Mycorrhizal fungi are particular about their host species. *Tricholoma caligatum* (top) grows only with hardwoods. The similar *Tricholoma magnivelare* (bottom) is the pine mushroom.

Endomycorrhizal fungi are probably the puppet masters of all land–based life on Earth.

given their cryptic nature and inability (for most species) to be cultured in the lab. Ironically, what is known about them is that they dominate the planet and are probably the puppet masters for all life on terrestrial Earth! (Although we cannot see many or even most of the species out there, we can track them by the DNA they leave behind in the soil. And they leave it everywhere.)

By far the largest group of endomycorrhizal fungi is the arbuscular mycorrhizal (AM) fungi (phylum Glomeromycota). Arbuscular mycorrhizas take their name from the arbuscules—highly branched structures that they form inside each root cell—where the exchange of water and nutrients occurs. Endomycorrhizal associations are found across a much broader array of plants than are EcM, including many plants growing in boggy, dry, or nutrient-poor soils. In fact, AM fungi are thought to be able to scavenge nutrients from the poorest of soils. Most plants, including grasses and cereals, vegetables, vines and bushes, and even those that do not form associations with EcM fungi, are known to partner with AM fungi.

One final note on mycorrhizal fungi: while many of our large charismatic mushrooms are saprotrophic, meaning they rot things such as woody or other organic debris, a significant number of our favorite mushrooms are ectomycorrhizal species, including amanitas, boletes, Russulas, Tricholomas, chanterelles, and even truffles. This means that, upon finding one that is seemingly arising from the forest floor, you know that it is, in fact, arising from the roots of its host tree. This also means that, to find prized edible species, experts seek out only specific host trees for their quarry.

RUSSULA DIVERSITY:
Russula species, shown on both pages, are found in just about any habitat worldwide. They have a brittle texture, have striate (or ribbed) cap margin, are exannulate (or ringless), and come in many colors.

RUSSULA IDENTIFICATION:
To determine particular species of
Russula, check to see if the cap has
a cracked surface, if the surface
can be peeled off, notable odors,
or a peppery taste when touched
to the tongue.

3

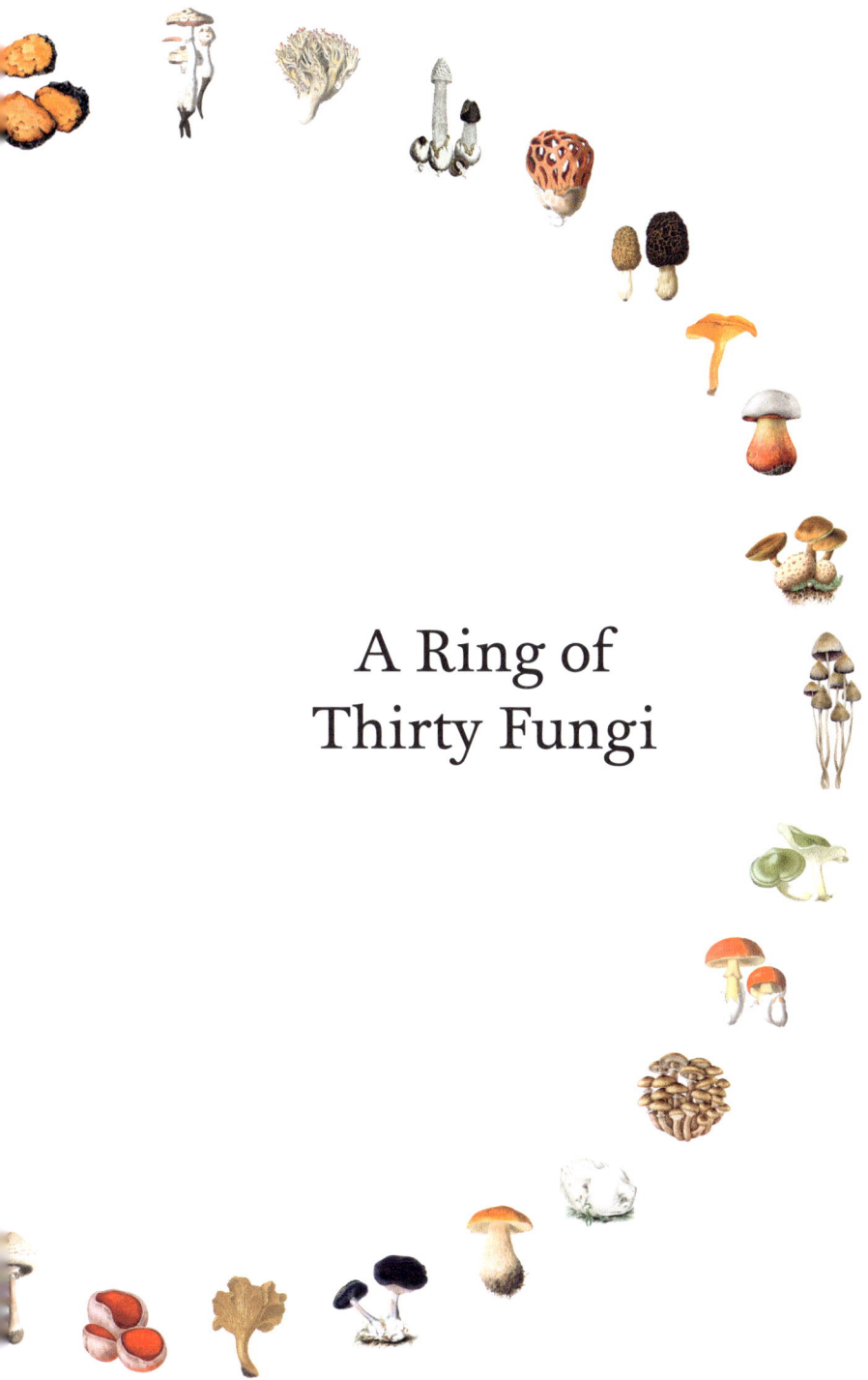

A Ring of
Thirty Fungi

Hebeloma syrjense & Hebeloma aminophilum

1 CORPSE FINDER & GHOUL FUNGUS

MORPHOGROUP
Agaric

TROPHIC MODE
Saprotroph

DISTRIBUTION
Northern Hemisphere;
Australia

HABITAT
Forests, urban areas

SUBSTRATE TYPE
Animal corpses,
nitrogenous waste

Many fungi are biotrophic and occur with a living host, while others decay organic matter (mostly plant material like dead grass, leaves, and wood). Now for something really strange! Some fungi are "sarcophilous"—keenly adapted to decompose animal carcasses or any organic matter that is highly nitrogenous or ammoniated.

Hebeloma syrjense.

A corpse may hardly be cold when the corpse finder (*Hebeloma syrjense*) and ghoul fungus (*Hebeloma aminophilum*) set to work. It is entirely likely that spores of these fungi are transported to corpses via flesh flies of the family Sarcophagidae, or other arthropods. These fungi are infrequently encountered and poorly understood, given their unusual ecologies.

What we do know of this group comes from the Japanese mycologist Naohiko Sagara, who made them his specialty. Fungi that are associated with nitrogenous matter include *Hebeloma*, which are mycorrhizal. So, on the one hand it's unusual that biotrophs also act as saprotrophs on nitrogenous waste or corpses of dead animals. On the other hand, since they seem to have a need for nitrogen in their diet (be it from a living tree host or other sources), it seems they have figured out a way to supplement this need. It is also likely that these few oddball sarcophilous species of *Hebeloma* truly are specialists and no longer partner with living trees. We just don't know.

Sarcophilous fungi often can be stimulated to fruit by burying urea or other compounds that decompose to ammonia in the woods. Thus, in the absence of a fresh corpse, other sources of ammonia are a suitable habitat. Sagara discovered that fruitings of *Hebeloma radicosum* could reliably be used to locate the dens of moles, where the mycelial source was the mammals' latrine. *Hebeloma sarcophyllum*, *H. syrjense*, and *H. radicosum* are uncommon but found across the Northern Hemisphere with *H. aminophilum,* known from Australia. They all grow in association with decomposing animal remains and sometimes are mentioned in literature for their utility in forensic science.

Ramaria botrytis

2 PINK-TIPPED CORAL

MORPHOGROUP
Coralloid

TROPHIC MODE
Biotroph, mycorrhizal

DISTRIBUTION
Northern Hemisphere

HABITAT
Forests

SUBSTRATE TYPE
Soil

Coral fungi look very much like undersea creatures, usually growing saprobically on (decaying) woody debris and logs, but many arise from soil. While some are drab, many are brightly colored (red, yellow, orange, and purple). They are indeed some of the most surprisingly pretty mushrooms of the forest.

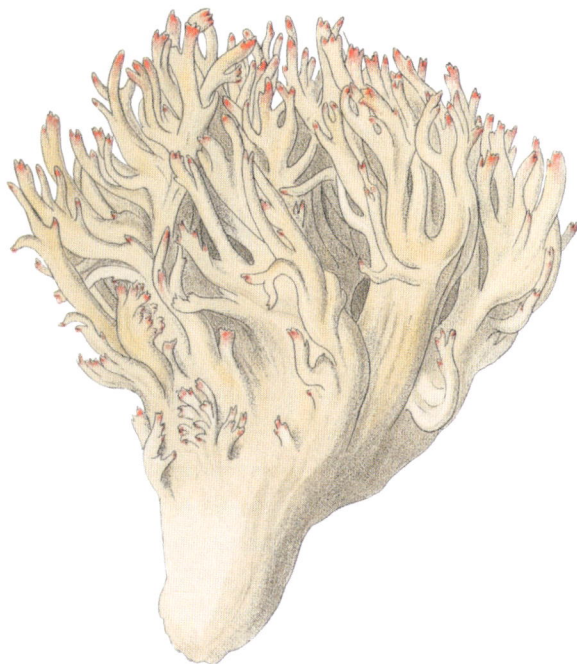

The first thing anyone notices is that coral fungi don't look like mushrooms. Their fruit bodies can be highly branching, or barely so—or entirely unbranched as with the worm corals. Their size is variable as well, ranging from a few inches tall and wide, to massive clusters more than 12 inches in diameter. Although there are no caps, gills, or pores, they do produce spores, which are borne directly from the side of branching stalks.

Coral fungi are numerous and commonly encountered, but they can be difficult to identify to a particular species. Possibly the most beautiful of all, and known from many diverse habitats, is *Ramaria botrytis*. This large, showy coral is considered an edible species, but many don't recommend collecting any corals for the table: they're mostly bland and with all those branches, nearly impossible to rid of grit and other forest debris. (Or maybe you enjoy "trail spice" in your dinner?) Moreover, many who have consumed this species complain of gastric upset and of . . . shall we say, laxative effects.

This species was first described scientifically in 1797 by the Dutch mycologist Christiaan Hendrik Persoon. It forms mycorrhizal associations with broad-leaved trees, especially beech, and fruits on the ground in wooded areas. You'll typically find it growing alone, scattered or in groups, under hardwoods or conifers during summer and fall.

This fungus has been the subject of study as the fruit bodies contain several chemical compounds with in vitro biological activity and antimicrobial activity against several species and strains of drug-resistant bacteria that cause disease in humans.

Phallus impudicus

3 IMPUDENT STINKHORN

MORPHOGROUP
Gasteroid

TROPHIC MODE
Saprotroph

DISTRIBUTION
Northern Hemisphere

HABITAT
Forests, urban areas

SUBSTRATE TYPE
Organic debris

Stinkhorns might be the strangest of all mushrooms. Some bear a slight resemblance to morels, but are basidiomycete mushrooms such as puffballs, gilled mushrooms, and boletes. Stinkhorns start out in an enclosed, egg-like form, then arise from a cuplike volva and produce a slimy, foul-smelling mass of spores over some part of the mature fruit body.

Stinkhorns are known on all continents except Antarctica, and many are very common urban mushrooms, living saprobically in organic debris. They seem to do just fine in anthropogenic settings; indeed, many have been introduced outside their native range through the importation of wood mulch and horticultural plants.

Where stinkhorns will turn up is unpredictable, but wherever spotted they always attract attention. Some resemble animals, such as squids or polyps. Others look like animal body parts, like arms or claws. This brings us to *Phallus impudicus*, which unabashedly resembles . . . animal genitalia.

Each of these large impudent stinkhorns has a white stalk. Rather than a cap with gills or pores at the top, there may be a bulbous tip (depending on the species), where a slimy, grayish-green, foul-smelling mass of spores ("gleba") is produced. This attracts flies, which may ingest or simply pick up the spores on their legs and bodies and carry them to distant locations.

Phallus impudicus (and the similar *Phallus hadriani*) have convoluted caps, somewhat like a morel. They grow to 6 inches or more and range across North America and Europe. *Phallus ravenelii* is similar but with a smooth cap and purple rhizomorphs attached to immature eggs from which another generation of stinkhorns will burst forth. Others are more or less prurient: the elegant dog stinkhorn, *Mutinus elegans*, has no cap—the slimy gleba is borne at the tip of the orange to pinkish-red stalk. The dog stinkhorn, *Mutinus caninus*, is similar but with a whitish stalk. Both grow to 6 inches and range across North America and Europe. Slenderer still, and with a small cap, is the Devil's dipstick, *Phallus rubicundus*, also found across North America and Europe.

Clathrus ruber

4 CAGE STINKHORN

MORPHOGROUP
Gasteroid

TROPHIC MODE
Saprotroph

DISTRIBUTION
Northern Hemisphere

HABITAT
Forests, urban areas

SUBSTRATE TYPE
Organic debris

No matter their mature shape, all stinkhorns start out in an enclosed, egg-like form. At first glance, you might suspect it to be a clutch of bird's eggs partially buried in organic debris. Or maybe some odd puffballs. Return to inspect them in a day or two and . . . what are those? *And what's that smell?* When mature, the "eggs" split open and an obscene fruit body pops out. (Stranger still: despite their off-putting smell, stinkhorns are consumed while in the egg stage by some adventurous mycophiles!)

While some stinkhorns bear a resemblance to morels or other mushrooms, many others come in bizarre forms, hardly resembling mushrooms at all. One that looks like a garish Wiffle ball, commonly seen in the west, southeast, and in much of southern Europe, is the cage stinkhorn, *Clathrus ruber*.

Most basidiomycete fungi produce and eject spores by ballistospory (see page 56). Not stinkhorns—they have taken a page from the book of flowering plants. Stinkhorns entice insect vectors to spread their spores in a less haphazard manner than releasing them to the vagaries of wind currents. In situations where fungal spores are transported on the exteriors of animal vectors, this is known as ectozoochory.

Most stinkhorns can be classified into two main groups within the order Phallales: those that are unbranched (and typically phallic-shaped, see page 98) and those that are branched. Branched stinkhorns may have the appearance of having "arms" or "claws" or they may appear as a "cage," as in the case of the beautiful *Clathrus ruber*.

With the Phallales, the mushroom produces putrid odors to attract scavenging flies. The (reportedly) sweet-tasting gleba spore masses are lapped up and stick to the flies who then transport the spores to other suitably unsavory substrates such as manure or rotting vegetation. Many flies even ingest the glebal goo—there is evidence that viable spores can pass through the digestive tracts of some fly species, as well. Rotting matter gives off odors that include chemicals called primary amines, which are enticing to all sorts of scavenging and carrion-feeding animals. Stinkhorns produce similar odors, so it's no wonder that scavenging flies are so strongly attracted . . . and that we humans are so strongly repulsed.

Morchella

5 MOREL

MORPHOGROUP
Cup fungi

TROPHIC MODE
Saprotroph, biotroph

DISTRIBUTION
Worldwide

HABITAT
Forests

SUBSTRATE TYPE
Organic debris

Morels are the most popular of wild edible mushrooms. Besides truffles, no wild mushrooms are so highly prized for their culinary value. And for most who pursue these oftentimes elusive springtime ascomycetes, it's a passion. There are books and movies about them, websites and clubs devoted to their pursuit, and all manner of tchotchke and other memorabilia featuring morel mushroom motifs.

All continents except Antarctica have morel species, and where there are morels, there are impassioned pickers who guard the secret spots where these springtime gems are collected. No wild mushroom is as enigmatic, has so much lore, and has been as bragged—and lied—about as the morel.

As elusive as the many black and yellow morel species are, there is another kind that is even more enigmatic—the mysterious burn morel. Also called fire morels, this kind of mushroom, known from Europe and North America, fruits only in the spring following a forest fire. In habitats that have never burned, they're hardly known.

No one really knows what sets them off, but in that first spring following a wildfire, the charred barren forest, reliably, will be carpeted with an eruption of morel fruit bodies that has to be seen to be believed. Researchers have postulated that changes in soil pH, salinity, or the release of nutrients after a fire somehow stimulate the mycelium to fruit. Or maybe the soil biology, chemistry, or microbial competitors change following the fire event. Recent studies suggest that the fungus is endophytic within badly burned—even killed— host plants, and then goes into reproductive mode to move on to pristine habitats by way of spores.

Morel mushrooms have now become big business, worth many millions of dollars. Over the past few decades, commercial pickers have been keying in on burn morels. Wherever there is a forest fire event, circle that place on your map and wait. Come spring, the burn morels will return to have their time to sporulate. But only for a few weeks ... then back into hiding they go to await the next big burn.

Omphalotus illudens

6   JACK-O'-LANTERN

MORPHOGROUP
Agaric

TROPHIC MODE
Saprotroph

DISTRIBUTION
Northern Hemisphere

HABITAT
Forests

SUBSTRATE TYPE
Wood

The jack-o'-lantern is a bright orange to brown-orange mushroom found growing from stumps or buried wood. It grows in clusters with stems fused at the base, with caps up to 8 inches wide and stalks 12 inches long.

All *Omphalotus* species are poisonous and produce a toxin called illudin that causes severe gastric upset. jack-o'-lanterns have a superficial resemblance to chanterelles (see page 152), and beginners do occasionally mix them up.

Omphalotus illudens is the bright orange jack-o'-lantern species from eastern North America and Europe (sometimes erroneously called *O. olearius*, but that species is European and mostly found on the wood of olive trees, as the name implies). *Omphalotus olivascens* is the species known from the West Coast, featuring a dark orange color with olive green tints.

Distinguishing this group from chanterelles is relatively easy. The gills of *Omphalotus* are orange and relatively sharp-edged (as opposed to the blunt ridges of chanterelles). The interior flesh of the jack-o'-lantern is orange, whereas the flesh of chanterelles is white.

The jack-o'-lantern is named for its pumpkin color and the fact that it is found in October, but also because the gills and other parts produce an eerie glow in the dark. That's right—these are bioluminescent fungi (see page 82). Bioluminescence in nature has been known and documented about since ancient times. Aristotle and Pliny the Elder both wrote about it.

There are just four known lineages of bioluminescent basidiomycete fungi, with around 80 different species known to do this. Mushrooms familiar to us that glow include *Armillaria*, *Mycena*, *Omphalotus*, and *Panellus*. If the light comes from hyphae in wood, often called "foxfire," it is most likely a species of *Armillaria*. The chemical basis for bioluminescence is well understood; the "why?" is a little tougher. That it serves in the attraction of invertebrates for the purposes of spore dispersal has been studied but doesn't seem to be the case in most habitats. Bioluminescence may simply be a way for fungi to dissipate energy as a by-product of oxidative metabolism or it may be tied to detoxification of peroxides that are formed during ligninolysis by wood-rot fungi.

Rubroboletus (Boletus) satanas

7   SATAN'S BOLETE

MORPHOGROUP
Bolete

TROPHIC MODE
Biotroph, mycorrhizal

DISTRIBUTION
Europe

HABITAT
Forests

SUBSTRATE TYPE
Soil

There are two deadly bolete species that occur in North America and at least one in Europe. These are known collectively as Satan's boletes. The "true" Satan's bolete, *Rubroboletus satanas*, is likely restricted to Europe. This mushroom has a white cap with a striking red stalk and pores.

There are myriad bolete species and many of them are pursued as choice edible wild mushrooms. If you are not proficient with bolete identification, you must be careful of some that bruise a blue color. Indeed, a safe bet is to avoid blue-staining boletes with red pores.

Rubroboletus satanas, formerly known as *Boletus satanas*, was first described by the German mycologist Harald Othmar Lenz in 1831. In the delightful book *Mushrooms & Toadstools* (1953), author John Ramsbottom describes how Lenz observed several reports of adverse reactions from people who had consumed this fungus (and even felt ill himself from vaporous "emanations" he had inhaled while studying this mushroom). Was Lenz simply paranoid? Possibly. But there is no denying that this mushroom is indeed poisonous. This led him to name it *Boletus satanas* as it was in his estimation that this was the most poisonous mushroom of all.

In North America, similar species known as *Rubroboletus eastwoodiae* and *R. pulcherrimus* also may be called Satan's (or Devil's) bolete. All share the characteristic red pores, red stems with yellow to blood-red reticulation, and light (white or nearly so) colored caps. All turn blue upon bruising or cutting, and all can be chunky and squat; typically, more broad than tall. All three are mycorrhizal with hardwood trees, primarily oak and beech. Two final notes: while all three Satan's boletes are among the most beautiful of bolete species, they are best admired in the forest. A good rule of thumb when it comes to picking boletes for the table: if they stain blue and the pores are reddish . . . you could wind up dead-ish!

Pseudoboletus parasiticus

8   PARASITIC BOLETE

MORPHOGROUP
Bolete

TROPHIC MODE
Biotroph, parasite

DISTRIBUTION
Northern Hemisphere

HABITAT
Forests

SUBSTRATE TYPE
Living fungi

Referred to as the parasitic bolete, *Pseudoboletus parasiticus* seems to be an obligate parasite on the poison pigskin puffball, *Scleroderma citrinum*. Maybe. Some think there may be more going on here than we know. Either way, the host is a very commonly seen false puffball. The parasitic bolete? Not so much. So, keep a lookout . . . if spotted, alert your mycophile friends. This one is on everyone's watch list.

With a few exceptions, boletes are well known as biotrophic fungi. That is, they form mycorrhizal symbionts with living partners—trees, mostly, and shrubs (see page 112). However, there are exceptions, and it's unlikely you have ever heard of them—and even less likely you know about the fascinating lifestyles of parasitic boletes.

No one really knows much about parasitic boletes, and they are only now being studied. What we do know is that within the very large family Boletaceae there are two small (and very strange) sister groups with very different and curious lifestyles. The trophic modes (are they saprotrophic? Biotrophic mutualists? Biotrophic parasites?) for the subfamily Chalciporoideae and the genus *Pseudoboletus* are unclear. They don't seem to be saprotrophs and are probably not mutualistic biotrophs either. It has been suggested that they are mycoparasitic, thus fungal parasites of other fungi.

Subfamily Chalciporoideae is the earliest diverging lineage in the Boletaceae and includes two genera: *Chalciporus* and *Buchwaldoboletus*. Both groups are weird. *Buchwaldoboletus* species fruit on wood and were thought to be saprotrophs, but the type species *Buchwaldoboletus lignicola* is now thought to be mycoparasitic on the wood decay fungus *Phaeolus schweinitzii*, the Dyer's Polypore. *Chalciporus* species are cute little boletes frequently seen fruiting at the same time and place as *Amanita muscaria*, the famous fly agaric. So, it was long thought that they were mycorrhizal friends, simply sharing the same host trees. Recent investigations suggest it may be more complicated than merely friends with benefits. It's more likely that *Chalciporus* species are parasites on *Amanita muscaria*. But the jury is still out; stay tuned as this story unfolds!

Psilocybe semilanceata

9 LIBERTY CAP

MORPHOGROUP
Agaric

TROPHIC MODE
Saprotroph

DISTRIBUTION
Worldwide

HABITAT
Grassy areas

SUBSTRATE TYPE
Organic debris

The genus *Psilocybe* is a large group (nearly 400 species worldwide) of small, brown mushrooms that grow saprobically on decaying organic material or the dung of mammals. While probably the best known of the genus, *Psilocybe cubensis* is not nearly so widespread as the cosmopolitan *Psilocybe semilanceata*, known as the liberty cap. This small brown mushroom has a bell-shaped cap with a sort of nipple ("umbo") at the top. Native to northern Europe, it commonly fruits on lawns and pastures all around the world.

Psilocybe species have very dark, mostly purple-black spore prints. Besides these, there is little about the group to distinguish it from other little brown mushrooms out there growing saprobically—for which there are many hundreds of species. In fact, there are so many found all over the world that mycologists simply call them all colloquially "LBMs." But this particular group are no mere LBMs.

What's so magic about *Psilocybe* mushrooms? The "active" species of *Psilocybe* (and a few closely related genera) contain psychotropic tryptamine alkaloid compounds (psilocybin, psilocin, baeocystin, or various combinations). All parts of the mushrooms seem to have these compounds, except for the spores. Tryptamines seem to function like serotonin in the brain, binding directly with one serotonin receptor in particular, and likely cause hallucinations by disrupting the integration of information coming from these sensory organs.

Psilocybin and psilocin have a very low toxicity (about 1.5 times that of caffeine) and do not seem to be addictive. Psychedelics like psilocybin and psilocin were used to successfully treat tens of thousands of patients in the 1950s and '60s and have recently returned to the forefront of research. These compounds have a profound effect on the mind and show much promise as therapeutic drugs.

Among psychedelics, psilocybin was recently shown to relieve depression symptoms rapidly, with sustained benefits for several months—all after a single dose of the drug. One of the outcomes of psilocybin is that it seems to "rewire" the brains of patients. This may be more than a metaphor. Researchers are finding that the therapeutic effects may arise from the ability of psilocybin to induce brain dendrite growth, both in culture and in animal models.

Clitocybe odora

10 ANISEED MUSHROOM

MORPHOGROUP
Agaric

TROPHIC MODE
Saprotroph

DISTRIBUTION
Northern Hemisphere

HABITAT
Forests

SUBSTRATE TYPE
Organic debris

Many members of the *Clitocybe* genus are rather plain looking and medium sized, with 2- to 4-inch caps. They have few unique features to separate them from each other; lacking a ring and volva, and rarely having bright colors. In most cases, their spore prints are pretty ordinary too—usually white, though some may feature pinkish or yellowish tints.

Some groups of mushrooms are difficult to identify; yet others you could almost ID with your eyes closed. All it takes is one whiff! Watch any expert field mycologist when they go to work on a mushroom. They will go through a checklist of characteristics, asking questions such as: Where was this collected and in which habitat? What sort of trees were in the vicinity? Substrate? Inevitably ... here it comes ... the sniff! (And maybe even a taste, too.)

A frequent forest encounter will be various members of the genus *Clitocybe*. For most, this is a tough group to identify. All clitocybes are saprobic, commonly found on decaying wood, leaf, or other forest litter. They can be found in hardwood or conifer forests, often in great numbers—sometimes rings.

Clitocybes do have one character that assists in recognizing them from other mushrooms: odors. Many are unpleasant, but some are quite fragrant—one especially so: *Clitocybe odora*. It has a smell strongly resembling anise (some say there are notes of vanilla, too). Scientists have even figured out the source of the aroma, a chemical called *p*-anisaldehyde (with a touch of benzaldehyde). This is one mushroom that I can positively identify, even with my eyes closed!

Other common clitocybes to learn to recognize include the massive, foul-smelling *Clitocybe nebularis*, *Clitocybe* (*Ampuloclitocybe*) *clavipes,* with its funnel shape and club-shaped, spongy stem; and the poisonous *Clitocybe dealbata*. Some clitocybes (e.g., *C. dealbata*) contain muscarine and—while probably not lethal—can still cause serious poisoning, potentially requiring hospitalization. For this reason, clitocybes should only be collected for the table by experts.

Amanita caesarea

11   CAESAR'S MUSHROOM

MORPHOGROUP
Agaric

TROPHIC MODE
Biotroph, mycorrhizal

DISTRIBUTION
Europe

HABITAT
Forests

SUBSTRATE TYPE
Soil

Caesar's mushrooms are some of the most photogenic members of the genus *Amanita*, and they are among the most prized of edible mushrooms in places where they occur. Despite the notoriety that amanitas have for being deadly, the Caesars are known to be some of the safest edibles within the genus. Indeed, many members in this section are sold in the open markets in Europe, Asia, and even Mexico.

Previously thought to be a single cosmopolitan species, Caesar's mushroom is now considered a group of similar-looking species numbering 83 in total, worldwide. *Amanita caesarea*, the first named of all the Caesar's mushrooms, is known from southern Europe and the Mediterranean, southward into northern Africa. The American Caesar's mushroom, *Amanita jacksonii*, known from eastern North America, looks similar. *Amanita hemibapha* is known from Asia, and there are other Caesar's mushrooms with red-orange caps, colorful stalks and gills, and striking white volva from elsewhere in the world. Indeed, there are others in this group with similar statures, much paler in color.

One of the most popular edible mushrooms in Southeast Asia is a very pale (cream to white), robust Caesar called *Amanita princeps*. Immigrants migrating to North America frequently mistake this favorite from back home with very similar look-alike amanitas . . . the deadly destroying angel, *Amanita bisporigera*, as well as the death cap—and all too often with fatal results. Deadly amanitas account for 90 to 95 percent of all mushroom-poisoning fatalities each year, globally.

Possibly the strangest story in the history of mycology and the most infamous of all murders attributed to poisonous mushrooms may involve this species. It is claimed that the death of the Roman ruler Claudius in AD 54 resulted from his fondness for *ovuli*, a famous Italian dish that utilizes this mushroom. Legend has it that Claudius was murdered by his fourth wife, Agrippina, who adulterated his beloved dish with death cap mushrooms so that her son Nero would take the throne. Following the murder, Nero did become the Roman ruler . . . and we know how that turned out.

Armillaria mellea

12 HONEY MUSHROOM

MORPHOGROUP
Agaric

TROPHIC MODE
Biotroph, parasite

DISTRIBUTION
Worldwide

HABITAT
Forests, urban areas

SUBSTRATE TYPE
Wood, living or dead

Armillaria mellea is known as the honey mushroom, but can go by additional common names, including shoestring or bootlace mushrooms. A popular edible mushroom (for some people) in North America and Europe, the honey mushroom can frustrate as it is highly variable in size, shape, and color. Modern science has shown that this variability in appearance is because the honey mushroom is actually a group of species of *Armillaria*.

One of the key identifying features of *Armillaria mellea* is the presence of tufts of dark hair near the center of the cap, and the fibrous nature of the stem. Another notable character is color (*mellea* for "honey-colored"): the true honey mushroom has a cap that is yellow-brown and a yellowish partial veil that leaves a ring on the stem.

Armillaria tabescens, and almost indistinguishable *Armillaria gallica*, have ephemeral cobwebby veils that leave no ring on the stem and are called ringless honey mushrooms. The latter two are usually orange-brown, and like their ringed relative, have tufts of dark hair near the center of the cap, a fibrous stem, and rhizomorphs. Species of *Armillaria* grow on wood, especially on oak and maple; another species, *Armillaria ostoyae,* grows on conifers.

Besides being extremely common in the fall—often in massive fruitings at the base of trees—these mushrooms are notorious for one additional reason. *Armillaria* species cause root and butt rot of living trees, then continue to utilize their dead host saprotrophically. These fungi spread by way of rhizomorphs ("shoestrings") growing on dead and dying trees and fallen logs. Rhizomorphs are darkly colored hardened strands of hyphae and can grow for many feet (or miles!) along the forest floor. (The largest-known organism on Earth is a honey mushroom mycelium that has covered more than 2,000 acres, or 800 hectares, in Oregon!) The dark color is attributed to melanin (the same compound that darkens our skin to protect from damaging sunlight). In the case of these parasitic fungi, their melanized rhizomorphs allow them to spread from tree to tree, felling one tree after another as they spread.

Calvatia gigantea

13 GIANT PUFFBALL

MORPHOGROUP
Gasteroid

TROPHIC MODE
Saprotroph

DISTRIBUTION
Worldwide

HABITAT
Grassy areas

SUBSTRATE TYPE
Organic debris

The surface of the giant puffball, *Calvatia gigantea* (also called *Langermannia gigantea*), is smooth and white when fresh, becoming olive to brown when mature. This fruit body is round, and when mature the surface splits open to release the spores. *Lots* of spores—a single specimen may produce upward of eight trillion! Specimens can be basketball sized or larger, and they may even fruit in fairy rings.

Giant puffballs are truly a sight to behold. On the one hand they can be surprisingly large—too large to hold with a single hand. On the other hand, they are well known to mycophiles and collected by many as safe, edible wild mushrooms. (What they lack in flavor and texture, they make up for with size.) Indeed, they are so safe that they are considered one of the "foolproof four" species of mushrooms that most guidebooks recommend to beginners wanting to gather wild mushrooms for the table.

Giant puffballs are frequently encountered in dry or exposed habitats, with sparse vegetation and grasses, and where frequented by grazing mammals. This style of fruit body is adapted to dry areas, as they keep their hymenium sequestered away inside and safe from drying out.

It is hard to miss a giant puffball, given their size . . . and just as difficult to resist kicking them! It has been suggested that this is part of their evolutionary adaptation to life on prairies; that they may benefit from being kicked or trampled by the hooves of mammals passing through while grazing.

This species is widespread and common across eastern North America and Europe. The giant puffball of western North America is *Calvatia booniana*, noted by its more elongated shape and surface that cracks and peels. The beautiful purple-spored puffball, *Calvatia cyathaformis*, can be quite small or nearly as large as the giant puffball, but has a spore mass that is purplish at maturity.

Boletus edulis

14 KING BOLETE

MORPHOGROUP
Bolete

TROPHIC MODE
Biotroph, mycorrhizal

DISTRIBUTION
Europe

HABITAT
Forests

SUBSTRATE TYPE
Soil

A fleshy mushroom, featuring a stalk with a cap that lacks gills, but has a layer of tubes instead. These tubes are arranged as a surface of pores from where spores are released into the air. Boletes make up an enormous group of macrofungi; nearly all are mycorrhizal symbionts of trees and other plants. Many dozen genera and hundreds of species inhabit every forest in the world.

One of the most highly prized of all edible mushrooms is the king bolete. Each fall, their return to the forests of Europe is eagerly anticipated. Italians especially go crazy for them; check out just about any produce stand or gourmet restaurant menu and you are sure to see them prominently boasted. Depending on where you are in North America, these mushrooms may appear in spring, summer, or fall. They are the stars of the show at the largest wild mushroom festival in North America, held each August in Telluride, Colorado.

The king bolete goes by dozens of other common names around the world, including porcino, cep, steinpilz, and penny bun. While their brownish caps (some species can be very light or very dark, almost black) are puffy and resemble baked buns, their flesh is firm, especially when young, and they keep their texture when cooked gently. If there is a finer-tasting mushroom, I cannot think of one!

We now think this is actually a group of many similar species. *Boletus edulis* of Europe is the most widely known, but there are at least a dozen others that are nearly identical in taste, texture, and appearance, found throughout North America and Asia, as well as Europe. All species within this group share a set of common physical features. They are large mushrooms, sometimes growing to weights of 3 pounds and reaching diameters of 15 inches. Often, the stem is more robust than the cap. They all have white pores when young (but may turn an olive yellow-green as they mature), and the key feature is that near the top of the stem, you will see a fine reticulation —a net-like pattern of raised tissue that is always white with king boletes.

Cortinarius violaceus

15 VIOLET WEBCAP

MORPHOGROUP
Agaric

TROPHIC MODE
Biotroph, mycorrhizal

DISTRIBUTION
Northern Hemisphere

HABITAT
Forests

SUBSTRATE TYPE
Soil

All species of *Cortinarius* are mycorrhizal and they will always be found in the presence of trees. There are many color variations within the genus. Brown and red are common, but some come in dazzling shades of purple— *Cortinarius violaceus* is a striking deep purple webcap of the Northern Hemisphere. *Cortinarius* could be the largest genus of gilled mushrooms on the planet. Though difficult to identify to a particular species, identifying a "cort" (or webcap) to genus can be quite easy.

Webcaps are medium-to-large mushrooms with 4- to 8-inch caps and rust-brown spore prints. They are most well known for their cortina—a cobwebby veil that protects the developing gills. As the mushroom draws in water and the cap expands, the cortina stretches and breaks, but, unlike many other agarics, no ring will remain on the stem. Many of the fibers will cling to the stem—invisible—until they start to collect the rusty spores as they drop.

Cortinarius species in general are not collected to be eaten. Though there are many safe edible species, there are a few poisonous ones, and some known to be lethal when consumed. In any case, most of the corts that are edible are not viewed as tasty enough to be worth the trouble—or risk. Furthermore, all mushrooms with rust-brown-colored spore prints (including the deadly *Galerinas*) should be avoided by everyone except experts in mushroom identification.

This is definitely a group to try to become familiar with. *Cortinarius* species seem to turn up just about anywhere there are forest trees. As one example, the far-ranging *Cortinarius magellanicus* is a mycorrhizal symbiont of *Nothofagus* trees native to the Southern Hemisphere (and typically fruits in late fall e.g., March, April, May), often in large groups. Even if not foraging for the table, *Cortinarius* is one of those genera best enjoyed for their beauty. And beautiful they are! From the garishly blood-red *Cortinarius sanguineus* to the delicately colored lilac *Cortinarius iodes*; the slimy *Cortinarius trivialis* and even slimier *Cortinarius vanduzerensis*; they are all a feast—for the eyes!

Cantharellus cibarius

16 CHANTERELLE

MORPHOGROUP
Agaric

TROPHIC MODE
Biotroph, mycorrhizal

DISTRIBUTION
Europe

HABITAT
Forests

SUBSTRATE TYPE
Soil

Chanterelles are highly popular edible mushrooms wherever they occur. Most species are yellowish in color, but they may also be white, pink, purple, red, brown, or black. All are vase-shaped and mycorrhizal, most often with oak species, but some partner with conifers. They range from about 1 to 4 inches in size, but some can be much larger.

There are two genera in the chanterelle group: species of genus *Cantharellus* are solid, including the stem; species of genus *Craterellus* are hollow or tube-like, including the stem. The gills are usually no more than small ridges that may run from the cap's margin down the stalk. Some are completely smooth with no gills whatsoever.

The most common yellow chanterelle of eastern North America is *Cantharellus flavus*, from western North America is *Cantharellus californicus*, and from the southeast is *Cantharellus tenuithrix*. All are fairly large, gregarious species that occur with oaks. Another yellow species, recognizable by its peach-colored tones, is the rainbow chanterelle, *Cantharellus roseocanus*; it ranges from the West into the Midwest. The Smooth Chanterelle (*Cantharellus lateritius*) has a nearly smooth spore-bearing surface.

The black trumpet (*Craterellus cornucopioides*) is highly prized across North America and Europe. When dried, they have an aroma of truffles or Romano cheese; in France they are known as "poor man's truffles." *Craterellus fallax* is similar and can be distinguished only by their spore print (*Cr. fallax* has salmon buff spores while *Cr. cornucopioides* spores are white). There are small species of *Craterellus*, some of them commonly seen, including the yellow-foot chanterelle group, *Craterellus ignicolor* and *Cr. lutescens* of Europe. Cute little *Craterellus tubaeformis* has a typically gray hymenium.

Urnula species look similar to black trumpets but are found in early spring. Chanterelles could be confused with the toxic jack-o'-lantern mushroom, *Omphalotus* species, which has a similar funnel shape, color, habit, and season. The false chanterelle, *Hygrophoropsis aurantiaca*, is similar in appearance, though it has true gills that fork repeatedly.

Sarcoscypha coccinea

17 SCARLET ELF CUP

MORPHOGROUP
Cup

TROPHIC MODE
Saprotroph

DISTRIBUTION
Worldwide

HABITAT
Forests

SUBSTRATE TYPE
Dead wood

There are numerous small cup fungi that you may encounter on fallen logs or other debris, or even bare soil. Many are drab shades of brown, but some may be quite colorful. *Sarcoscypha* species are very commonly seen; these small, bright red cups (1–2 inches in diameter) show up between late fall and spring on partially buried wood. *Sarcoscypha coccinea*, the scarlet elf cup, is one such species.

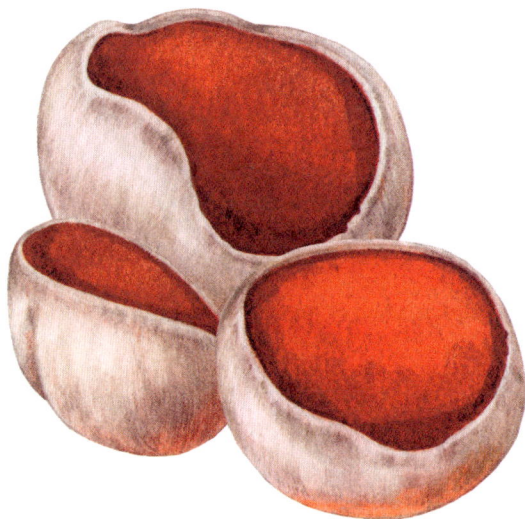

Sarcoscyphas are always attached to decaying wood, by a short or sometimes very long stalk, depending on the species. Their cup- or saucer-shaped fruit body (apothecium) is smooth on both surfaces. The Scarlet Elf Cup is one of several fungi whose fruit bodies may be observed (or heard) to produce a "puffing" sound, as thousands of asci simultaneously explode to release a smokey cloud of spores!

The scarlet elf cup has many look-alikes. *Scutellinia* species (eyelash cups) are frequently seen on very moist, mossy, well-rotted logs or stumps, or along forest streams. Their flattened cups have dark, bristly hairs resembling eyelashes around the cup edge, but you may need a hand lens to see them. Species of *Cookeina* look like hairy little funnels on the forest floor or rotting logs. They range in color from pink to bright red, some are even yellowish. Many, like *Cookeina tricholoma,* may seem quite bristly or hairy; the specific epithet *tricholoma* has to do with "hair" and "fringe." If you are a mycophile in the Northern Hemisphere, undoubtedly these species are on your "life list" to see at some point. When encountered, their apothecia are often filled with water.

Cookeina was named in honor of Mordecai Cubitt Cooke (1825–1914), an English botanist and mycologist. *Cookeina tricholoma* is a member of the family Sarcoscyphaceae (order Pezizales). Besides *Sarcoscypha coccinea* and *Cookeina tricholoma*, this family includes other peculiar fungi, ranging from tiny cups on delicate stalks (*Microstoma*) to massive globose ones resembling witches' cauldrons of the genus *Sarcoscypha*. All species of *Cookeina* occur in the tropical and subtropical regions of the world.

Chlorophyllum molybdites

18 GREEN GILL

MORPHOGROUP
Agaric

TROPHIC MODE
Saprotroph

DISTRIBUTION
Worldwide

HABITAT
Grassy, urban areas

SUBSTRATE TYPE
Soil

Our only mushroom with green spores, the green-gilled lepiota is a big white mushroom that often appears in large fairy rings on lawns in warm and especially hot climates. In the middle of a hot summer, a passing downpour can result in a fairy ring of huge, beautiful white mushrooms the very next day or so.

Parasol and shaggy parasol mushrooms are mostly found growing out in the open, often in parks or near dwellings. They favor mulch and leaf litter, as well as grassy areas. All are saprobes and rot grassy or woody debris. This group includes some of the largest stalked mushrooms, exceeding 12 inches in height and 10 inches in cap diameter. You may hear parasols and shaggy parasols referred to colloquially as "lepiotas" because they were once considered species of the genus *Lepiota*.

All parasols are white-spored and feature a stalk with ring and gills that remain white at maturity, with one very notable exception: a false parasol, *Chlorophyllum molybdites*. Also known as the green-gilled lepiota, this species is quite toxic and is one of the most common causes of mushroom poisonings worldwide. This mushroom is common in the tropics, subtropics, on islands, and throughout the world wherever it gets hot and where there are grassy, open areas. It is even spotted infrequently in northern North America, when conditions are right.

While the true parasols and shaggy parasols are famously prize edibles, you must not confuse the green gill for them. The sickness that ensues will be remembered for the rest of your life! Note that the absence of green spores is not foolproof: the gills will turn greenish gray only once the specimens are very mature or over the hill.

Russula virescens

19 QUILTED GREEN RUSSULA

MORPHOGROUP
Agaric

TROPHIC MODE
Biotroph, mycorrhizal

DISTRIBUTION
Worldwide

HABITAT
Forests

SUBSTRATE TYPE
Soil

Russulas do not have large, showy individuals to draw in photographers. They have a white to yellow spore print, attached gills, and typically no ring on the stalk. They are closely related to the milk mushrooms, *Lactarius* species, which have similar characteristics but emit a milky latex when their gills are broken. Microscopically, this genus is characterized similarly by amyloid ornamented spores.

Known as the brittle gills, this species of *Russula* comprises a large genus of medium-sized (2–4 inches wide) terrestrial mushrooms. They are on every forest floor across the globe but don't get the attention that other groups of mushrooms get. Their bright colors catch the eye of the beginner but the fascination fades as their limited value is discovered.

Quilted green russula are famously fragile—some species can barely be handled without breaking. Why so brittle? Species of *Russula*, *Lactarius*, and other closely related groups (of the order Russulales) feature hyphae composed of spherocysts— cells that are dramatically inflated, with thin cell walls. These cells will easily break, shattering the mushroom, unlike most other kinds of mushrooms that are flexible.

Russula is a large genus composed of around 750 worldwide species of ectomycorrhizal mushrooms. Anyone familiar with mushrooms will recognize the brittle gills. They are common in woodlands and even urban areas, fairly large, and brightly colored—often in striking red shades, although they may be golden, brown, or purple-black. To my eyes, the most beautiful are the green and blue-green species like *Russula virescens* and some of the better *Russula cyanoxantha*.

They are not often sought out as table fare. Edible Russulas can be found, but typically they are of mediocre flavor and poor texture; many taste like hot peppers or have an otherwise acrid flavor. None of the Russulas in Europe or North America are deadly poisonous, but many, such as *Russula emetica* of Europe and its North American look-alike *Russula silvicola*, can cause gastric upset. Both are beautiful mushrooms with a bright red cap and very white gills and stem. They have an incredibly hot taste, which confirms that they are not a palatable species.

Hericium erinaceus

20 LION'S MANE

MORPHOGROUP
Tooth-mushroom

TROPHIC MODE
Saprotroph

DISTRIBUTION
Worldwide

HABITAT
Forests

SUBSTRATE TYPE
Dead wood

The lion's mane or bearded tooth (*Hericium erinaceus*) is a large, white, stalkless tooth fungus found on broadleaf logs, stumps, and living trees. It consists of one large clump of teeth, with the basidia borne on the outer surface. In addition to being commonly found in nature, this is one of the most cultivated mushrooms in the Northern Hemisphere.

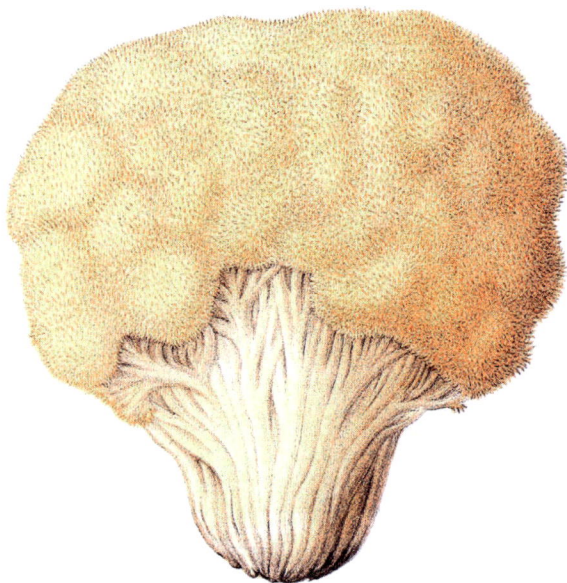

Mushrooms without a stem (astipitate), featuring teeth or spines (rather than gills or pores) are tough to group. They aren't all closely related; rather they have evolved similar morphological features because of convergent evolution. This results from adapting to a certain niche in the environment, where the ecology drives the morphology of the organism—much to the frustration of scientists!

Such is the case with members of the order Russulales, a large group of mushroom genera that share a common ancestor. You might not be surprised to learn that *Russula* and *Lactarius* are closely related, but would you believe that species of *Hericium* are just as closely related to both? *Hericium* species produce spores borne on teeth or spines that hang from the underside of the cap, rather than gills, but the fruit body has no stalk. *Hericium* is characterized by amyloid ornamented spores and feature hyphae composed of spherocysts.

The similar bear's head tooth (*Hericium abietis*) grows from conifers, especially hemlock and fir. Coral tooth (*Hericium coralloides*) occurs on hardwoods, and has numerous clumps of teeth, about ½–¾ inch long. All range across North America and Europe. The comb tooth (*Hericium americanum*) has relatively short teeth (about ¼ inch long), which come from both sides of the branches and point in all directions. All look like pieces of white coral and are tasty edibles.

Hericiums have one large look-alike: the northern tooth (*Climacodon septentrionale*). This stalkless toothed mushroom grows on standing trees (mostly maples), with overlapping shelves seen high above ground. The teeth are tougher than other fleshy tooth fungi, like the *Hericium* species. Although very large and showy, it is inedible.

Leucocoprinus birnbaumii

21 FLOWER POT LEPIOTA

MORPHOGROUP
Agaric

TROPHIC MODE
Saprotroph

DISTRIBUTION
Worldwide

HABITAT
Forests, urban areas

SUBSTRATE TYPE
Organic debris

Older books may list this mushroom as *Lepiota lutea*, given its appearance and white spores. Many small lepiotas and lepiotoids resemble this one, along with the much larger parasols and shaggy parasols. Members of the small lepiotoid group typically are less conspicuous, mostly reside in forests, and are smaller (2–4 inches tall with caps about 2 inches in diameter).

The flower pot lepiota (*Leucocoprinus birnbaumii*) is a beautiful, albeit small, lemon-yellow mushroom found in mulch, flower beds, and grass. The mycelium is often present in nonsterile potting media, and the mushrooms subsequently develop in flowerpots and greenhouses. The upper part of the stem is thin and fragile, and the base is swollen. Although one of the most attractive mushrooms, this species is poisonous.

Lepiota magnispora and *L. clypeolaria* are quite showy and have shaggy stalks; they are found across North America and Europe and are all but indistinguishable. *Lepiota cristata* and *L. subincarnata* have a similarly wide range but feature smooth stems with similar colors as the reddish cap disc. *Lepiota lilacina* is smaller with lilac tones and seems to favor mulch in urban settings.

One other member of this group is worth mentioning. The small lepiotoids may be quite showy and several are very shaggy, including their stems. Caps typically are darker colored at the center. None should be considered edible; some contain the same deadly amatoxins that are found in some amanitas. All are saprobic in plant debris; the smooth lepiota (*Leucoagaricus naucinus*) occurs on lawns. All have white gills, a stalk with rings, and a somewhat bulbous stalk base resembling amanitas.

Coprinus comatus

22 SHAGGY MANE

MORPHOGROUP
Agaric

TROPHIC MODE
Saprotroph

DISTRIBUTION
Worldwide

HABITAT
Forests, urban areas

SUBSTRATE TYPE
Debris, disturbed areas

The shaggy mane or shaggy inkcap (*Coprinus comatus*) gets its name from the prominent woolly scales on its cap, reminiscent of an old lawyer's wig. The long stems are white, hollow, and feature a central cord. The annulus, if present, is fragile and low on the stem. The largest of the coprinoids, they sometimes reach heights of 16 inches or more.

Coprinus comatus can appear almost overnight and have a complete life cycle of just a few days. The gills rapidly mature then dissolve into an inky mess. Found on lawns, humus, and compost piles, they are an excellent edible, but you must get them early!

Coprinus comatus is one of only a handful of species in the genus *Coprinus*. *Coprinus* and *Coprinus*-like ("coprinoid") mushrooms are often called inky caps. The four genera that make up the coprinoids are: *Coprinus*, *Coprinellus*, *Coprinopsis*, and—the smallest and most fragile—*Parasola*. They are saprotrophic and nearly all may be found worldwide.

Mica cap (*Coprinellus micaceus*) gets its name from the film of crystal-like granules on the young caps. The orange-mat inky (*Coprinellus radians*) is so-called because of the fuzzy mat of sterile, orange-brown mycelium ("ozonium") that is produced on the woody substrate at the base of the cluster of mushrooms. It is found on damp wood on the forest floor and in basements.

The alcohol inky or tippler's bane (*Coprinopsis atramentaria*) produces coprine, a toxin that blocks the breakdown of alcohol in the body, causing unpleasant gastric distress and a full-body burning sensation—essentially a powerful hangover. Alcohol consumed up to five days after eating this mushroom could produce these symptoms. It is found at the base of dead trees, stumps, or from dead, buried wood, often in urban settings.

The magpie inky (*Coprinopsis picacea*) is strikingly pretty. Surprisingly tall for a coprinoid (at least 8 inches), this denizen of European woodlands and urban parks features large shaggy scales overtop a deep chocolate-brown cap. Equally pretty, though lighter brown, is the scaly ink cap (*Coprinopsis variegata*), which occurs on dead wood.

23 PARROT WAX CAP

MORPHOGROUP
Agaric

TROPHIC MODE
Biotroph, endophytic

DISTRIBUTION
Worldwide

HABITAT
Forests, urban areas

SUBSTRATE TYPE
Soil

These small, glossy mushrooms can range from bright yellow to deep crimson and feature every color of orange in between. Wax caps often grow in large troops, coloring the forest floor just before the leaves cover it in fall. They can be striking, especially the genus *Hygrocybe*, which includes the witches' hat (*Hygrocybe conica*), a red-orange mushroom that turns black upon handling.

Wax cap is a general name used for small forest and grassland mushrooms that have a waxy feel. At one time, all were grouped together in the genus *Hygrophorus*, then more recently that group was split in two, with *Hygrophorus* species considered biotrophic (always found with trees) and sister genus *Hygrocybe* seeming saprotrophic and usually occurring in grassy areas, often with no trees in sight.

That these colorful mushrooms are attractive cannot be denied—many are brightly colored—but looks can be deceiving. It turns out that while all the species of wax caps appear similar, science has determined that convergent evolution has been at work, and not all these species are close relatives. And so, what was once a single genus and then two genera, the group of wax caps are now members of several genera: *Cuphophyllus*, *Gliophorus*, *Humidicutis*, *Hygrocybe*, and *Hygrophorus* (although many books, and even some mycologists may still use only the genus *Hygrocybe*.)

Despite their small size (about 1 inch or so), the wax caps will still catch your eye. Possibly the most striking of all, and one of our few green mushrooms, is the parrot mushroom (*Gliophorus psittacinus*). What it lacks in size, it certainly makes up in beauty. Found over most of the world, it usually occurs in moist, mossy areas where it can blend in with green to yellow-green backgrounds, and it takes its name from its green color. This mushroom, when fresh, glistens with a wet, almost slimy covering. Depending on where you encounter this beauty, the green color may fade to other shades as it ages: sometimes yellow, orange, pink, or even blue.

Amanita phalloides

24 DEATH CAP

MORPHOGROUP
Agaric

TROPHIC MODE
Biotroph, mycorrhizal

DISTRIBUTION
Worldwide, distribution
increasing

HABITAT
Forests, urban areas

SUBSTRATE TYPE
Soil

Unfamiliar pickers erroneously assume that poisonous mushrooms will warn of impending danger with garish colors, foul odors, or off-putting tastes. In nature, most toxic or venomous animals and plants do display aposematic or "warning colorations"— bright red or yellow, for example—but fungi do not follow these rules. The most commonly encountered poisonous mushrooms are drably colored browns or grays; many are pure white.

One of the most widespread mushroom species worldwide, *Amanita phalloides* was first described from Europe and is now known on all continents, except Antarctica. We know more about its ecology than most mushrooms because wherever it turns up, death soon follows. Responsible for most mushroom-poisoning deaths (90 to 95 percent) worldwide, experts predict that the numbers of poisonings from death cap mushrooms will continue to rise.

That *Amanita phalloides* is now so widespread is attributed to its ability to pair up with a wide assortment of host trees, including horticultural and economically important nut, lumber, and pulpwood species, thus it is transported and transplanted globally. In North America, the death cap is associated with oaks (including coast live oak and tanoak), European beech and linden (in urban settings) on the West Coast, and various oaks on the East Coast. In just a few decades' time, the range of this mushroom has expanded dramatically and there is no reason to think it won't continue to grow. This mushroom has made the headlines in Australia following several deaths, including patrons who'd consumed wild harvested mushrooms at a restaurant. Europeans are wary of this species because it has always been there.

If you are at all interested in collecting wild mushrooms for food, you absolutely must be familiar with this and all deadly species of *Amanita* mushrooms. Dangerous mushrooms, including death caps, oftentimes resemble other familiar edible mushrooms, even cultivated species. Furthermore, most taste quite pleasant—there's nothing to warn you that what you're savoring in a prepared dish is about to kill you.

Amanita muscaria

25 EUROPEAN FLY AGARIC

MORPHOGROUP
Agaric

TROPHIC MODE
Biotroph, mycorrhizal

DISTRIBUTION
Europe, Asia,
distribution increasing

HABITAT
Forests, urban areas

SUBSTRATE TYPE
Soil

Amanita muscaria is a big mushroom, often with a cap that can be a foot in diameter atop a stalk that may be over a foot tall, with a scaly, bulbous base. Not all populations are the same and, despite being so well-known, the fly agaric is still very much a mystery. It is incredibly common and widespread—known on all continents except Antarctica.

This is without doubt the most recognizable mushroom worldwide. Whenever a mushroom is needed for illustrations, postcards, cartoons—even emojis—this handsome red mushroom with white scales is reproduced. It has been fancifully depicted for so long that it's sometimes easy to forget there is a mushroom that matches this description in nature.

The current scientific understanding is that there are multiple subspecies of *Amanita muscaria*. The original description came from the red variety of Europe and Asia; there is a different red variety in western North America, and eastern North America has a yellow variety. This mycorrhizal fungus partners with a range of tree hosts. Even the colors aren't absolute; red varieties can range from red to orange to yellow to cream, and yellow varieties vary across the spectrum too.

Recently, scientists have determined that the European fly agaric is an aggressive invasive fungus, now spreading all over the world. *Amanita muscaria* is found in Australia, New Zealand, Argentina, Brazil, Chile, and Tanzania. This mycorrhizal symbiont of trees seems to be moving around with pine and *Eucalyptus* plantation stock. Recently, it's made its way to North America—populations have been detected in Alaska, California, and Massachusetts.

Although this mushroom is good news to the lumber industry, promoting the growth of plantation trees outside of their native range, it doesn't seem to stay put and is jumping to native species in its new home. In North America, it is often found growing in stands of native birch trees. What this means for the future of forests is uncertain, but many fear the invasive European fly agaric may outcompete and push out native mycorrhizal fungi that could be key components to healthy ecosystems.

Ustilago maydis

26 CORN SMUT

MORPHOGROUP
Rusts and Smuts

TROPHIC MODE
Biotroph, parasite

DISTRIBUTION
North America, Europe

HABITAT
Agriculture, farmland

SUBSTRATE TYPE
Plants

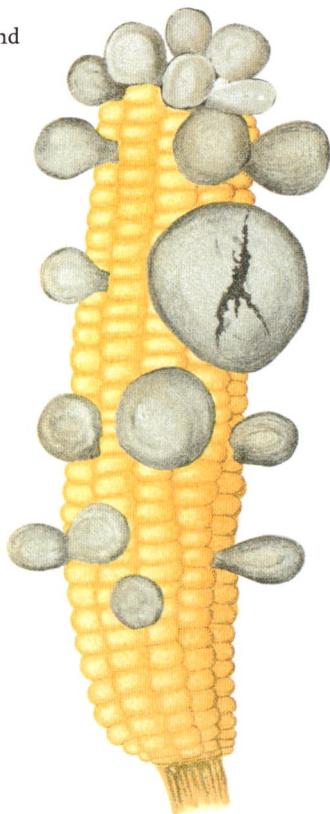

Ustilago maydis is edible and has long been considered a delicacy in Mexico, prepared in all manner of ways including in ice cream. (It tastes better than it looks, with flavors of mushroom, corn, chocolate, and vanilla.) Sometimes called "Mexican corn truffle," *huitlacoche* (also spelled *cuitlacoche*) was the Aztec name, which roughly translates to raven's "excrement" or "droppings." A personal favorite nickname for it is that of myco-raconteur David Arora: "porn on the cob."

Looking more like excrement than a mushroom, and with an unsavory name to match, corn smut is a conspicuous fungus with an amazing life cycle. This basidiomycete parasite of corn (maize) plants can be found throughout warmer North America and Europe. Historically, the fungus was common on field and sweet corn, but modern varieties are resistant. Heirloom corn is still susceptible, as is popcorn and Indian corn.

All parts of the plant may be infected, but galls are mostly seen on ears because the silk (an extension of the female part of the plant) is receptive to pollination—as well as fungal invasion. The life cycle of smut fungi features two spore stages. The large galls are a mass of black, sooty ("smutty") teliospores enclosed in a smooth covering of plant tissue. Teliospores overwinter, their germination timed to the reproduction cycle of the corn plant. Teliospores germinate in the soil, giving rise to hyphae with club-shaped basidia; borne on each are tiny basidiospores ("sporidia").

Haploid sporidia alight on corn plants but are not yet able to infect the host. First, they must germinate, growing in a yeast-like manner in search of a partner. Successful crossing between two mating types restores the dikaryotic condition. Armed with a full complement of genes, the smut fungus is now infectious—but still needs luck. If on the silk, the fungus must reach the ovary before pollination occurs. If the fungus lands anywhere else on the corn plant, it cannot penetrate the tough cuticle of the corn plant unless damaged (e.g., by hail or insects). Damage to plant tissues can facilitate infection via sporidial or telial hyphae. Thus, outbreaks of corn smut are frequently associated with episodes of hail damage.

Tuber melanosporum

27 PÉRIGORD TRUFFLE

MORPHOGROUP
Truffle

TROPHIC MODE
Biotroph, mycorrhizal

DISTRIBUTION
Europe

HABITAT
Forests

SUBSTRATE TYPE
Soil

The key to the truffle's success is its odor. Truffles are irresistible to humans; their aroma described as earthy, garlicky, musky, or sexy. The chemical most responsible, 2,4-dithiapentane, is industrially synthesized and used to create "truffle-flavored" oils and other foodstuffs. Counterfeiters, who have long been known to mix cheaper truffle species (including Chinese species and European summer truffles) into batches of Périgords, now adulterate batches with synthetic aromas.

Few fungi have the allure and mystique of the truffle. While many species of truffles are collected commercially from around the world, the Périgord truffle (*Tuber melanosporum*) of France and the Piedmont white truffle (*T. magnatum*) of Italy dominate the market and command far and away the highest prices (US$1,345–4,700 per kilogram on today's market). Demand far exceeds supply and wild-collected yields are notoriously unpredictable.

Everyone always asks: if they're so difficult to find in the wild, why not cultivate them? Truffle cultivation is notoriously difficult, in part due to a clandestine life cycle underground. Truffle fungi are mycorrhizal symbionts; the species mentioned above live on the roots of oak (*Quercus* spp.) and hazelnut trees (*Corylus avellana*). Their hyphae extend outward in all directions, and if they fuse with another of their kind, a fruit body may result. Ascospores are produced in the fruit bodies that remain underground, and thus rely on animals for dispersal. Mycophagous animals, including wild boars and rodents, dig up and eat truffles, then the spores pass through the digestive tract and are dispersed with feces. The truffles are located by smell; components of their aroma are irresistible mimics of mammalian sex pheromones.

In order to combat counterfeit truffles, biologists are working to create a complete genome of the Périgord truffle fungus. It is hoped that this will lead to rapid tests to determine the authenticity of all truffles to be used at the time of sale. Already, researchers are compiling a database of genetic markers to verify the geographic origins of truffle species from around the world.

Serpula lacrymans

28 DRY ROT FUNGUS

MORPHOGROUP
Resupinate, mold

TROPHIC MODE
Saprotroph

DISTRIBUTION
Worldwide

HABITAT
Urban areas, rarely forests

SUBSTRATE TYPE
Wood

When you think of all the calamities that cause damage to human dwellings, common molds probably do not come to mind. Hurricanes, tornadoes, floods, and fires all make the headlines, but the pervasive damage to buildings by molds and other fungi is still very real. Of all the wood decay fungi that damage timber constructions worldwide, the creepy-looking dry rot fungus is considered the most destructive and is one to be feared.

This cosmopolitan fungus wreaks destruction the world over. A recent study found that losses by UK building owners to ameliorate damage caused by this organism was estimated to be at least US$200 million annually. In the USA, the damage is much worse, estimated at US$17 billion annually.

Dry rot has likely been living with us ever since humans began creating dwellings from wood—it is even mentioned in the Bible. As humans spread around the globe, this fungus traveled right along, seemingly benefitting from humanity. Strangely, this common urban fungus is rarely found in nature—perhaps it doesn't compete with the myriad microbes in competition for the same carbohydrates of dead wood.

Serpula lacrymans causes brown rot primarily of conifer wood and is keenly adapted to life in the dried timbers of our homes. *Serpula* has the amazing ability to transport water by way of mycelial cords or rhizomorphs, often over great distances— even through foundations. The increase of water content in completely dry wood facilitates further colonization in unfavorably dry areas. Wood decomposition subsequently creates additional water as a by-product of fungal catabolism and respiration, acting as a feedback loop and further colonization. Anywhere wood is present, *Serpula* can show up. Although modern construction increasingly uses synthetic materials, this doesn't deter it. Interestingly, it can utilize quite a few inorganic materials for its nutritional needs, including calcium and iron ions extracted from plaster, brick, and stone.

It is hard to believe the closest relatives to dry rot are the majestic boletes. The Finnish mycologist Petter Karsten named it *Serpula* (Latin for "creeping") and *lacrymans* (a reference to the ever-present drops of liquid exudate that resemble tears.)

Inonotus obliquus

29 CHAGA

MORPHOGROUP
Resupinate

TROPHIC MODE
Biotroph, parasite

DISTRIBUTION
Northern Hemisphere

HABITAT
Forests

SUBSTRATE TYPE
Living wood

Chaga is produced by the polypore *Inonotus obliquus*, a white rot fungus in the family Hymenochaetaceae. It is a Northern Hemisphere species that grows primarily on birch trees, producing gnarled black tumors, often called "clinkers," for they resemble pieces of burned charcoal. The yellow-brown interior has a cork-like consistency.

Few Westerners had heard of chaga before Aleksandr Solzhenitsyn introduced it in his 1968 novel, *The Cancer Ward*. Solzhenitsyn was a Russian author and Soviet dissident who helped to raise global awareness of political repression in the Soviet Union, especially the Gulag prison system. The protagonist of his largely autobiographical novel, Oleg Kostoglotov, is a political prisoner who is diagnosed with cancer upon his release from the Gulag. An old country doctor treats him with a well-known (to the locals) cure for cancer based on a strange birch tree fungus—chaga.

It turns out that chaga has been used in Russian folk medicine since at least the sixteenth century. Indigenous people of Siberia used chaga as a tea to prevent cancer, and heart and liver disease. It was also smoked for respiratory ailments. It has since become popular with mainstream Russians who use it to treat "consumption" and cancers—often of the stomach and lungs. In 1955, a refined extract of the chaga fungus (called "Befungin") was developed by the Botanical Institute of Russian Academy of Sciences, and it continues to be included in the official state pharmacopoeia.

There has recently been a huge upsurge of interest in products made with chaga, including nutritional supplements, capsules, tinctures, elixirs, teas and coffee immune boosters, bottled beverages—and more everyday it seems—all for the purported health benefits of this fungus. Unfortunately, the chaga gold rush is starting to leave its mark. More and more birch forests are being cut down or severely damaged as harvesters try to keep up with a seemingly insatiable demand for this fungus. Will chaga go the way of the elephant, tiger, and rhino? Only time will tell.

Ganoderma lucidum & Ganoderma applanatum

30 LING CHIH & ARTIST'S CONK

MORPHOGROUP
Polypore

TROPHIC MODE
Saprotroph

DISTRIBUTION
Worldwide

HABITAT
Forests

SUBSTRATE TYPE
Wood

The conks are perennial and put down a new layer of tubes each year. Young conks might be less than an inch in diameter, while old specimens might be 2 feet across or more. They vary in color from ash gray to tan to brown. Copious amounts of brown spores may be deposited on any surfaces nearby.

Ganoderma lucidum.

Mushrooms are typically an ephemeral part of the fungus life cycle, appearing for just a few hours to days during favorable parts of a year. Many other mushrooms are more reliable; annual polypores typically sporulate one year then wither, although their old fruit bodies may persist before falling off the host tree. They are usually rubbery and pliant. Perennial polypores persist on their host tree for multiple years, putting on a fresh new hymenial layer on the underside of the fruit body each growing season. Typically hard and woody; if sliced in half, they even display growth rings.

One of the most common bracket fungi, the artist's conk (*Ganoderma applanatum*) gets its name from the fact that the fresh pore surface can be easily engraved with a knife, stick, or fingernail. It may be found on hardwood and conifers and is primarily a rotter of dead wood but will occasionally cause decay in living trees. This species has become popular for its purported medicinal properties. Historically, it was used for fiber. Conks were pounded into a surprisingly tough, waterproof, felt-like material, often used for hats in Europe.

The varnished conk (*Ganoderma tsugae*) is found almost exclusively on hemlock (*Tsuga* spp.) trees in eastern North America, while the Oregon artist's conk (*G. oregonense*) is found on conifers in the West. Ling chih (*Ganoderma lucidum*) is similar and occurs on hardwoods and some conifers, ranging across North America and Europe. The cap may be 12 inches across or more and colored a deep reddish brown to dark maroon; the growing edge is white, as is the underside. When fresh, these mushrooms are shiny and appear to be coated with varnish. Ling chih has long been used for purported medicinal properties in China and other parts of Asia.